TRANSACTIONS

OF THE

AMERICAN PHILOSOPHICAL SOCIETY

HELD AT PHILADELPHIA

FOR PROMOTING USEFUL KNOWLEDGE

NEW SERIES—VOLUME XXXI, PART II
FEBRUARY, 1939

CHRONIC ARTHRITIS IN WILD MAMMALS

HERBERT FOX

PROFESSOR OF COMPARATIVE PATHOLOGY, UNIVERSITY OF PENNSYLVANIA;
DIRECTOR, WILLIAM PEPPER LABORATORY, HOSPITAL OF THE
UNIVERSITY OF PENNSYLVANIA

PHILADELPHIA:

THE AMERICAN PHILOSOPHICAL SOCIETY

104 SOUTH FIFTH STREET

1939

LANCASTER PRESS, INC., LANCASTER, PA.

CHRONIC ARTHRITIS IN WILD MAMMALS

Being a Description of Lesions Found in the Collections of Several Museums and from a Pathological Service

HERBERT FOX

Professor of Comparative Pathology, University of Pennsylvania [1]

Pathologist, Philadelphia Zoological Society

(*Read in part at the General Meeting, April 21, 1938*)

ABSTRACT

The analysis of more than seventeen hundred skeletons and autopsies of wild animals reveals in the joints changes that correspond with chronic arthritis in man. Not only has this been discovered in specimens exhibited in menageries but also in material that was certainly in its proper wild habitat when killed. The lesions of these "truly wild" animals are entirely comparable to those from captive specimens. It is evident therefore that chronic arthritis occurs in nature.

It has not been difficult to discover these cases and it has been reasonably simple to learn which varieties have the most conspicuous lesions, namely anthropoid apes and baboons, Felidae, Hyaenidae, Ursidae, Bovidae, Cervidae and a few others. Reversely it has developed that a number of groups, notably certain families of Carnivora (e.g. Canidae) and some orders like the Rodentia and Chiroptera, do not appear to have arthritis in the material of this study although a very considerable number have been examined. The ease of discovery of the disease in hyaenas and gorillas should be emphasized.

No attempt is made to give percentage; it is only stated that 77 cases were accepted as arthritic in the study of 1,749 skeletons listed in Table I.

Arthritis is seen best as an involvement of the spinal column, but it occurs also very extensively in the appendicular skeleton.

The distribution of lesions in the different kinds of animals suggests that there may be a relationship of function and localization of disease, possibly related to locomotion and the jolt shock associated therewith.

Conspicuous arthritis-bearers are macrosomic animals; small bodied animals, rodents and bats, etc., are missing.

There is no apparent relationship between arthritis and taxonomic position, zoogeography, ecology, habits, diet, pathological panels and focal infection.

There is a strongly suggested similarity between the arthritis of the lower mammals and that of the deforming and rheumatoid arthritides in man.

CONTENTS

[1] From the Department of Comparative Pathology, University of Pennsylvania and the Penrose Research Laboratory of the Philadelphia Zoological Society. Aided by a grant from the Faculty Research Committee, University of Pennsylvania.

INTRODUCTION

Standing high in the list of ailments of mankind and one that incapacitates very great numbers of people, is chronic arthritis. Since knowledge of the nature and causation of this malady is so lamentably inadequate, the following study to increase our information is justified.

It is the purpose of the present article to put on record the results of investigation of arthritis as seen in the skeletal material from many museums and from the autopsy service at the Zoological Garden in Philadelphia. Such a survey seems not to have been made before. There are isolated examples of arthritis in the reports from the London Zoological Society and other zoological agencies, but the data are inadequate for combination with those in the following notes, which comprise an observational study of spontaneous pathology and apparently the first one attempting to cover the field.

Arthritis, as well as other pathological conditions, will probably vary in different classes of animals, but the results that would be most clearly applicable to man occur in the mammalian group. The evolution of practically every part of the body in the mammal, from monotreme to man, indicates that comparable demands of normal physiological anatomy have been met similarly at all stages. Thus the anterior shoulder girdle of man can be seen to have its homologue in the narrowly built, limited shoulder complex of the ungulate or the widespread, powerful equipment of the aard vark. If this general principle of normal evolution be accepted, it would appear consistent to assume that abnormal strains or assaults from within or without might be effective in a comparable, evolutionary manner and that these processes would be applicable to man.

While anatomical characters have followed evolutionary lines it cannot be proved as yet that physiological characters have had a parallel evolution. All mammals as a class have the same organs with similar functions but the physiological relationships are not fully known. It is hoped that this study may elucidate in part muscle and bone functions.

Nor is there true knowledge of the role of parasitism throughout the mammals; infectious diseases, of both animal and vegetable causation, occur in all members of the group but not to the same degree.

Evaluation of every possible influence that might be effective in producing arthritis in mammals demands a full review of the entire field, a requirement that cannot be fulfilled in this paper but can be started by a physical study of available material. To this end we are forced to use menagerie and expeditionary material. The domestic animals are excluded from this study for they have been cultivated by man so long that they have lost much of their comparative value. They represent but few of the zoological orders. They have much arthritis, particularly draught and circus horses, but these can hardly be compared to wild, odd-toed ungulates.[2]

SOURCE OF MATERIAL

The collection of material included in this study comprises the autopsies at the Philadelphia Zoological Garden (P.Z.G.) that have been studied closely since March 1, 1935, most of the bodies having been macerated and disarticulated, and also skeletons placed at our disposal at the following places:

Wistar Institute of Anatomy of Philadelphia (W.I.), by the late Milton J. Greenman, by the late H. H. Donaldson, and by the President, Dr. Alfred Stengel; the Departments of Zoology and Anthropology of the National Museum (N.M.) at Washington through the agency of Ales Hrdlicka and G. E. Miller; the Wagner Free Institute of Science (W.M.), Philadelphia, by the Director, Mr. Boyer; the Department of Anatomy and Haman Museum, Western Reserve University, Cleveland (Td.), T. Wingate Todd, Director, where the extensive collection of the Pongidae yielded much valuable material; the Academy of Natural Sciences, Philadelphia (A.N.S.), where the help of Witmer Stone and Wharton Huber was at our disposal; the American Museum of Natural History, New York City (A.M.), in the departments of W. K. Gregory and H. E. Anthony.

The writer is authorized by the University of Pennsylvania and the Zoological Society of Philadelphia to acknowledge this help and to express thanks for the privileges and courtesies received at these places. He also wishes to express personal appreciation for the fraternal attitude of scientific freemasonry shown by his hosts on the many visits that were made to the various institutions, here and in other cities. A detailed account of each group of skeletons studied at each museum seems hardly necessary to include here, but every specimen selected for individual description is identified as to its source by the initials of the above named institutions. A list of material examined, in the zoological classification recently adopted by the Zoological Society of London, is to be found in Table I. Both arthritic and non-arthritic mammals are included in these figures.

DEFINITIONS

The pathological lesions sought in this study can be included in the following definition. Chronic arthritis as known today is a malady of unknown cause, potentially polyarticular, with no proven relationship with any other co-existent or previous abnormality, that produces a permanent alteration of articular surfaces, synovia, cartilages, epiphyses, para-articular ligaments, capsule and supra-epiphyseal bone; not all of these parts may be affected to the same degree. There is a consequent limitation of motion and some degree of deformity. *Restitutio ad integram* probably never occurs when the process ceases, unless it be after the very mildest of attacks.

The names given to the principal forms by the human pathologist and clinician are legion and examples are given at this point for explanatory, not controversial, reasons. The more active form is called infectious or atrophic or rheumatoid, and is more often seen

[2] See Hare, 1927, Veterinary Record, 7: 411 and 432.

in the first half of life, while the characteristically less flagrant type is called hypertrophic or degenerative or osteo-arthritis. The controversy about their relation and causation is not entered into here, but it is hoped that some features of arthritis in wild animals may help to settle this question.

It is, however, necessary that the supposed sequence of events within joints be explained since many of the stages are found in the cases reported here. Arthritis is rarely recognized during the life of the wild animal. It can be suspected by various alterations in well known joints, such as the wrist complex of the hyena or the shoulder complex of the felines.

Fresh tissues of the autopsy table will show hyperplastic synovia, fringes, erosions of cartilage and ligaments, hyperplasia of epiphyseal edges as in man. While early stages, or at least less destructive stages, are limited to the parts making up the joint proper, later or more deforming lesions are to be found deep in the epiphysis and at the edges of epiphyseal junction with the diaphysis. Actually, inflammatory stages will be associated with congestive, exudative or infiltrative processes in the spongy bone, just beneath the sub-cartilaginous osseous cortex. As the inflammation advances, both the cartilage on the bone ends and the thin cortex may ulcerate. During the time occupied by the changing of the bone ends, the para-articular cartilages and synovial fringes swell by edema or infiltrate, the capsule proper becomes distended by exudate within it and thickened in itself by infiltrate, and the ligaments are involved in a manner similar to that in the capsule. If the process continue, bone ends soften and are so altered as to lose typical characters; the supporting tissues or ligaments do likewise. Should the process be slow, or should reparative efforts be exhibited by the tissues, granulation tissue may appear between articulating parts, fibrosis develop and healing by direct ankylosis result. Should the inflammation be greater around the joint proper than on the apposed surfaces or their immediately subjacent neighboring parts, then para-articular lesions would be greater.

While the foregoing applies principally to limb joints, similar changes occur in the vertebral joints. The mildest alteration appears to be an irregularity of the anterior edges of vertebral bodies, sometimes by erosion, at others by hyperostosis, so-called lipping. Then defects of lateral and posterior margins of the individual bones follow, with irregularities in the width and elastic character of the discs which may retract or ulcerate and even disappear. Sometimes changes appear on the articulating faces of the vertebrae, as poroses, while the cartilaginous discs remain in fairly good condition, but primary damage to the interosseous cortex of the vertebral units is always followed by degeneration of the cartilaginous tissue between them. Distortion or crushing of the vertebrae is frequent under this condition. Hyperostoses along the anterior and lateral surfaces of the bodies and of the apophyses and costo-vertebral joints are prominent and appear to originate always from joint lines.

The most important feature of this hyperplastic process seems to be that, at the joint line, density of bone by excess calcification can be found, by cutting and by x-ray, but *within* the body of the bone, rarefaction can always be found.

Whereas certain cases of arthritis appear to begin in bone ends or on terminal hyaline cartilages, leading to intra-articular necrosis, fibrosis and ankylosis, while other cases exist in which the joint surfaces are less damaged but peri-epiphyseal and ligamentous inflammation is prominent and leads to fibrosis and ossification so that false external ankylosis results, many students have insisted that the two forms are different in etiology and prognosis and they support their claims by clinical features of chronic arthritis in man.

The lesions found in the material to be analyzed were studied with this basic anatomico-pathological process in mind and when, in the following, any name or pathological term is used, the definition accepted by pathologists as describing the physical object—bone, joint, etc.—is intended, not a human clinico-pathological diagnosis. It would indeed demand an entry into the contention to explain how wild material conforms to one or other variety so that it will not be attempted. However, the basic changes of early vascular distention and distortion, the early infiltration into synovia and capsule, the early and late erosions and necrosis of cartilage either on bone ends or lateral thereto, the later reaction to cessation of inflammation or incident to repair, fibrosis, cicatrization and calcification—all are essentially the same in man and lower wild animals. Rate and intensity of the pathological process might explain the differences of the results. The cause of localization for rate and intensity might be conditioned by anatomy.

It has been possible to add many röntgen ray pictures of the bare bony specimens. The cooperation of Dr. T. W. Todd of Cleveland must be emphasized, for he supplied, in his laboratory and at his own expense, many photographs and radiographs of the Primates.

Many radiographs and direct photographs, often of the same specimens, are reproduced with simple notation of the readings. For the most part they have been submitted to trained röntgenologists and the readings are in accord with their interpretations. Evidence of hyperostotic overgrowth, osseous solidification, porosis and surface defects appear to be comparable to those observed by similar methods in examination of human joints.

Mention can be made of the degree to which radiographs contribute to diagnosis by reference to certain individual cases. Attention may be drawn to the thoracic spine of the Leche antelope P.Z.G. 9,592 illustrating the delicacy of the osseous overgrowth on its anterior surface and around posterior articulations, as in contrast to the heavier shadows in the bear P.Z.G. 11,800. Combined hypertrophic and ulcerative processes seem to be shown very well in the case of the leopard W.I. 3,681. The most suggestive bony and articular changes in the spine appear in the mandrill P.Z.G. 12,147 where all forms of disease and deformity appear.

The photographs of the gorilla material are stimulating, for all varieties of arthritis in man are suggested. Everywhere in the body of these animals one finds osteoarticular disease. That in Td. 1,991 suggests inflammatory arthritis and osteitis with secondary hypertrophy. The case Td. 1,731 suggests very early changes in the bone ends of infective nature that may resemble the earliest stages of rheumatoid arthritis.

Whatever the proper nomenclature to be given these changes, there appears to be little doubt that chronic arthritis can be found in wild animals by radiology, supporting the interpretation of the gross examination.

LIMITATIONS IN THE MATERIAL

A word as to our evaluation of the material seems proper before it is described in detail. Specimens that come from menageries have been subjected to the conditions of captivity for varying lengths of time, have been fed variously and enclosed in shelters of various sorts. Wild specimens were taken by naturalists on expeditions and by hunters. Naturally only those specimens could be taken in the wild that could come into the open

TABLE I

LIST OF ANIMAL SKELETONS AND AUTOPSIES EXAMINED, ARRANGED ACCORDING TO ZOOLOGICAL
CLASSIFICATION, WITH COMMON NAMES OF PRINCIPAL EXAMPLES

	Complete Skeleton	Incomplete Skeleton	P.Z.G. Aut.	Total
Primates				
Pongidae—Gorilla—gorilla..................	69	21	0	90
Pan—chimpanzee............. ..	23	0	1	24
Pongo—orang utan..............	41	0	0	41
Hylobatidae				
Hylobates—gibbon...................	22	0	0	22
Cercopithecidae				
Pithecus—entellus, etc.................	6	0	1	7
Cercopithecus—green, grivet, etc........	5	0	4	9
Cercocebus—mangabey................	2	0	3	5
Macaca—macaque....................	26	0	10	36
Cynopithecus—black ape...............	1	0	0	1
Theropithecus—gelada baboon..........	3	0	0	3
Papio—baboon.....................	39	0	2	41
Mandrillus—mandrill.................	2	0	1	3
Cebidae				
Saimiri—squirrel monkey..............	1	0	1	2
Cebus—capuchins.....................	14	0	4	18
Lagothrix—woolly monkeys.............	1	0	0	1
Ateles—spider monkeys...............	0	6	0	6
Alouatta—howlers....................	8	0	0	8
Hapalidae				
Hapale—marmosets..................	3	0	0	3
Lemuridae—lemurs..................	9	4	0	13
Lorisidae—loris, potto................	9	0	0	9
Galagidae—bush-baby...............	1	4	3	8
Daubentoniidae—aye-aye.............	1	0	0	1
Tarsiidae..........................	4	0	0	4
Unidentified monkeys.................	34	0	0	34
				389
Menotyphla				
Macroscelididae—elephant shrew........	16	0	0	16
				16
Lipotyphla				
Erinaceidae—hedgehog...............	8	0	0	8
Soricidae—shrews....................	13	3	0	16
Talpidae—European mole..............	23	0	0	23
Solenodontidae—solenodon.............	4	0	0	4
Centetidae—tenrec...................	2	0	0	2
				53
Dermoptera				
Galeopithecidae—flying lemurs.........	5	0	0	5
				5

TABLE I (*Continued*)

	Complete Skeleton	Incomplete Skeleton	P.Z.G. Aut.	Total
Chiroptera				
Pteropodidae—fruit-bat..............	2	0	0	2
Vespertilionidae—bats ⎫				
Emballonuridae—bats ⎬	68	0	4	72
Molossidae—bats ⎭				
				74
Carnivora				
Felidae—cats........................	112	1	9	122
Cryptoproctidae—fossa...............	0	1	0	1
Viverridae—civets, genets, paradoxures,				
ichneumons.....................	42	0	5	47
Protelidae—aard wolf.................	1	0	0	1
Hyaenidae—hyena...................	14	3	1	18
Canidae—dogs, wolves, foxes, jackals,				
etc............................	46	5	7	58
Mustelidae—marten, skunk, weasel,				
badger.........................	38	0	7	45
Procyonidae—raccoon, bassaris, coati,				
kinkajou........................	54	0	14	68
Ailuridae—panda....................	1	12	0	13
Ursidae—bears......................	24	2	8	34
				407
Pinnipedia				
Otariidae—eared seal, sea lion..........	3	0	3	6
Odobenidae—walrus..................	1	0	0	1
Phocidae—common seal...............	16	0	2	18
				25
Cetacea				
Delphinidae—porpoise................	1	1	0	2
				2
Rodentia				
Sciuridae—squirrels, spermophiles, mar-				
mots...........................	14	6	17	37
Castoridae—beaver....................	4	0	4	8
Muridae—rats, mice..................	10	0	7	17
Geomyidae—gophers.................	4	0	0	4
Bathyergidae—mole rat...............	1	0	0	1
Aplodontidae.......................	1	0	0	1
Jaculidae—jerboa....................	1	0	0	1
Zapodidae—jumping mice..............	2	0	0	2
Heteromyidae—kangaroo rats..........	1	0	0	1
Hystricidae—porcupine (Asiatic and				
African)........................	24	0	5	29
Octodontidae—Chilian bush rat........	2	0	0	2

TABLE I (*Continued*)

	Complete Skeleton	Incomplete Skeleton	P.Z.G. Aut.	Total
Capromyidae—hutia................	1	0	0	1
Myocastoridae—coypu rat............	2	0	0	2
Dinomyidae—paca-rana..............	3	0	0	3
Cuniculidae—cavy..................	5	4	4	13
Dasyproctidae—agouti................	11	0	3	14
Vizcaciidae—chinchilla, viscacha........	1	0	0	1
Hydrochoeridae—capybara.............	5	0	0	5
Unidentified rodents.................	31	0	0	31
				173
Lagomorpha				
Leporidae—hare, rabbit...............	6	0	0	6
				6
Proboscidea				
Elephantidae—elephants..............	4	0	1	5
				5
Hyracoidea				
Procaviidae—hyrax..................	8	0	0	8
				8
Perissodactyla				
Equidae—horse, ass..................	18	1	2	21
Rhinocerotidae—rhinoceros............	13	0	0	13
Tapiridae—tapir....................	19	1	1	21
				55
Artiodactyla				
Bovidae—oxen, antelopes, sheep, goats, etc.............................	59	2	26	87
Antilocapridae—pronghorned antelope...	17	0	1	18
Giraffidae—giraffe..................	11	0	2	13
Cervidae—deer, moose, elk............	36	5	26	67
Tragulidae—chevrotain, mouse deer.....	5	0	0	5
Camelidae—camels, llama.............	14	1	1	16
Suidae—swine, wart hogs.............	17	0	1	18
Tayassuidae—peccaries...............	21	0	0	21
Hippopotamidae—hippopotamus........	3	2	C	5
				250
Sirenia				
Trichechidae—manatee...............	3	0	0	3
				3
Tubulidentata				
Orycteropodidae—African ant-bear, aard vark...........................	20	0	0	20
				20

TABLE I (*Continued*)

	Complete Skeleton	Incomplete Skeleton	P.Z.G. Aut.	Total
Pholidota				
Manidae—pangolin..................	4	0	0	4
				4
Xenarthra				
Choloepodidae—two-toed sloth.........	5	3	0	8
Bradypodidae—three-toed sloth........	20	7	0	27
Myrmecophagidae—anteater..........	40	3	2	45
Dasypodidae—armadillo	20	0	2	22
				102
Marsupialia				
Macropodidae—*Macropus*—kangaroo, wallaroo, wallaby............	16	0	3	19
Petrogale—rock wallaby...........	11	0	0	11
Aepyprymnus—rat kangaroo.......	1	1	0	2
Potorous—rat kangaroo...........	2	0	0	2
Phalangeridae—*Trichosurus*—long-eared opossum, short-eared opossum...	0	0	2	2
Phalanger—cuscus................	4	0	0	4
Phascolarctidae—*Phascolarctos*—koala, native bear..................	3	0	0	3
Phascolomiidae—*Phascolomis*—wombat..	2	0	0	2
Peramelidae—*Perameles*—long-nosed bandicoot...................	2	0	0	2
Dasyuridae—*Thylacinus*—Tasmanian wolf........................	1	0	0	1
Sarcophilus—Tasmanian devil......	1	0	0	1
Dasyurus—viverrine native cat......	4	0	0	4
Didelphiidae—*Didelphis*—American opossum.......................	5	0	19	24
Unidentified marsupials..............	....	....	....	64
				141
Monotremata				
Tachyglossidae—Australian spiny anteater or echidna..............	4	4	0	8
Platypus.......................	3	0	0	3
				11
Grand total.................	....	....	....	1,749

and then presumably fully competent to look after themselves. Weak ones would hide or could not get about well. Poor specimens though taken in the field might be discarded and thus only good ones would be brought to museums. Dr. Todd's material of the Primates is largely from missionaries who purchased the bones from natives who are supposed to kill for food; such animals are probably to be considered average wild beasts.

Despite these limitations, one has to use the material available. It is interesting that arthritis acceptable for this study was found in truly wild animals and in menagerie specimens to about the same degree. However, no attempt shall be made to give percentages or probable incidence on so small a number as 1,749 examples of skeletons showing 77 cases of arthritis; this is a report of what was found, with very cautious deductions therefrom.

ANATOMICAL CHARACTERS OF ARTICULAR CHANGES IN MAMMALS, ARRANGED IN ORDER OF ZOOLOGICAL CLASSIFICATION

MARSUPIALIA

Kangaroo. A.M.

The first three cervicals are normal. The lower face of the 4th shows inferior lipping and porosity on the right side with slight eburnation toward the midline. There has been right dislocation and eburnation on the right corresponding to the above dislocation. The lower face of the 5th is normal. The upper face of the 6th is normal. The lower part of the body and margin of the 6th shows distinct lipping corresponding to superior lipping of the 7th. There is no eburnation.

Hypertrophic spondylitis with dislocation, porosity, eburnation of 4th and 5th cervicals. Hypertrophic spondylitis of the 6th and 7th.

Wallaby. A.M. 3,837

The 12th thoracic shows roughening along the anterior border of the body without corresponding roughening of the upper edge of the 13th. The lower edge of the 13th is rough and turned upward on the right side by a large irregular excrescence on the 1st lumbar. This 1st lumbar has an ossification on its lower left side which is probably attached to the intervertebral disc. The upper edge of the 2nd lumbar is rough and irregular but not greatly damaged. The 2nd and 3rd have false ankylosis and over-riding lipping. The condition of the 3rd lumbar is not clear.

This is a moderate grade of hypertrophic spondylitis of the last thoracic and upper three lumbars. The rest of the skeleton appears to be normal.

Kangaroo (*Macropus giganteus*).   ♂   P.Z.G. 12,193

In menagerie exhibition for 66 months; death from unexplained pulmonary hemorrhage. Skull thoroughly broken. Mandible, including teeth, apparently normal. Cervical vertebrae normal above 6th and 7th which show hyperostosis and eburnation on the ventral surfaces of the bodies and erosion on the left half of the articulating surface, eburnation on the right half, which indicates that the intervertebral disc had disappeared, the bones had been apposed and direct friction had occurred. The vertebrae are normal as far as the third lumbar which shows marked inferior dentoid lipping and disappearance of the anterior half of the vertebral base. The location of the nucleus pulposus is normal. The anterior and posterior extremities and the pelvis are all normal.

PROBOSCIDEA

Elephant (*Elephas indicus*). ♂ P.Z.G. 1,461

Results of original autopsy: Hypertrophic and degenerative spondylitis lower cervical and 3rd lumbar vertebrae. Wild born male and on exhibition 235 months. Had been noted as " rheumatic " and losing flesh for some time. Died without special history. Chronic myocarditis, chronic nephritis, chronic ulcerative tuberculosis of the lungs and lymph nodes, perilobular cirrhosis of the liver and pigmentation of the spleen. The joints were swollen for the most part and in the right hind leg second joint there was an especially large accumulation of clear fluid. The tip end of the right femur was ulcerated at the edge where the cartilage joins the bone. In all the carpal and tarsal joints and the articulations of these with the phalanges the cartilages were irregular and hard. There were no calcareous deposits.

Inspection of the skeleton some years later revealed the following: The epiphyseal ends of the bone all over the body show an abnormal porosity. This is more developed around the ends of the bones and not evident on the direct articulating surface. The edges of the ilium and scapula are especially rough and the condyles of the mandibles are uneven. The epiphyseal border of the metatarsal bones are unusually rough. The articulating surfaces are in some places very smooth as if eburnated, in other places ridged and rubbed as if the cartilage had been damaged at that point. The vertebrae show thoracic productive osteitis (4th and 5th; 9th and 10th; 18th and 19th). The osteophytes of the 9th and 10th are approaching each other as if ankylosis were about to be produced. The articulating surface of the right humero-ulnar joint appears to be the most eroded. Changes in the left humero-ulnar joint approach it.

UNGULATA

PERISSODACTYLA

Equidae

Wild Horse (*Equus caballus*). ♀ P.Z.G. 11,959

In menagerie exhibition 261 months. Skull is normal. There is no evidence of caries, but there is medium wearing down of all the teeth.

The backbone shows throughout varying grades of osteoporosis and great fragility. The spine as a whole is light in weight. The only joint at which changes specific of arthritis are found is between the 6th and 7th cervicals where the concavity of the 6th is deepened in proportion to adjacent cavities and the depth is irregular as if the cartilage had disappeared, there being some holes into the spongy portion of the body of the bone. The hemispherical dome of the 7th articulating surface is osteoporotic and very irregular. This constitutes an *ulcerative arthritis of the articulating surfaces but there is no periarthritis.* There is slight lipping of the 7th, 11th and 12th vertebrae on the inferior edges. The rest of the vertebrae, the pelvis, the scapulae, the extremity bones are all within normal limits. This animal has ulcerative spondylitis, atrophy and decalcification of the thorax and spine, and probably hypercalcification of the extremities. Thyroid and parathyroid normal.

TAPIRIDAE

Tapir (*Tapirus terrestris*).   ♀   P.Z.G. 12,160

In menagerie exhibition 124 months. Death from endometritis and enteritis. The skull is normal and the teeth are in unusually good condition. The 5th cervical shows inferior lipping without corresponding lipping in the 6th. The 1st thoracic shows anterior thickening at the location of the fibers of the longus colli. From the 3rd to the 12th thoracic there is slight antero-inferior lipping with very slight lipping of the superior surfaces of the next vertebrae.

It would appear that ossification was taking place in the anterior ligament and also in the intervertebral disc—especially between 5th and 6th thoracic. The 14th and 15th are tightly ankylosed, apparently because of complete false ankylosis. There probably was some wedging of both of these vertebrae because the length of these two combined is definitely less than that of the two above or below. The remainder of the vertebrae and all the bones are normal.

GIRAFFIDAE

Giraffe. A.M.

The 8th, 9th and 10th thoracics show superior and inferior lipping, the greatest prominence of which seems to follow the limits of the longitudinal ligament. The remainder of the skeleton is within normal limits.

HIPPOPOTAMIDAE

Hippopotamus (*Hippopotamus amphibius*). N.M. 162,979. Shot wild. Theodore Roosevelt Expedition

The only lesion discoverable is in the sacro-iliac region and the last three lumbar vertebrae. The articulating surface of the ilium at the sacral edge seems almost uninvolved but the articulating edge of the sacrum, the facet of the articulation with the last three lumbars, show hyperostosis without involvement of the articulating surfaces proper. The upper vertebrae seem normal.

Chronic hypertrophic osteoarthritis of the last lumbars and sacrum.

ARTIODACTYLA

SUIDAE

Wart Hog (*Phacochoerus* sp.). A.N.S. 1,355

Bi-lateral osteosclerotic and ulcerative lesions of both hip joints with degeneration and mushrooming of both capites femorum. Slight periosteal growth about the condyles of the femur. Nothing below. Anterior extremity negative. The thoracic vertebrae are in a bad state of preservation but appear to have hyperosteosis on the ventral surfaces and lateral processes with lipping and there was probably an ulceration of the intervertebral surfaces.

Wart Hog (*Phacochoerus* sp.). A.N.S. 12,930

The thoracic spine shows hyperostosis and lipping of ventral surfaces but apparently no ulceration of the articulating surfaces. No ankylosis. Osteoarthritis of the left humero-ulnar joint with such great overgrowth that the bones cannot be separated after cleaning. No subluxation but slight wearing down of the humerus. Corresponding phalanges show similar changes of periostotic character without change in the articulating surface. Left elbow good specimen.

Wild Boar (*Sus scrofa*). ♂ P.Z.G. 11,974

In menagerie exhibition 194 months. Death from senility, hepatic and renal disease. Teeth much worn and broken; no obvious caries. Mandible—osteomyelitis behind the molars of a type suggesting actinomycotic disease.

In the vertebral column there are no changes down to the 10th thoracic whereafter lipping of both upper and lower edges begins. It is most marked between the 1st and 2nd, 2nd and 3rd and 3rd and 4th lumbars. The posterior sections of the vertebrae are within normal limits for an old animal. The intervertebral faces begin to show roughening and exposure of spongy bone from the 9th downward so that the faces of the 1st and 2nd lumbars are entirely worn away. The lower face of the 2nd lumbar is within normal limits, as are the remainder. The rib sockets are all roughened as if there had been low-grade periosteal overgrowth, but the vertebral and cartilaginous ends do not show any overgrowth.

It should be noted that all the larger bones of this animal—mandible, ilium, ischium, scapulae—show considerable ridging, as if muscular insertions had pulled upon the periosteum and produced overgrowth. This is done in a rather orderly and symmetric fashion and is therefore interpreted as senile periosteal overgrowth, possibly a traumatic or tension reaction.

Scapulae are within normal limits, as are also the humeri, ulnae and radii. Both femora show slight thickening over the pre-trochanteric face of the lower neck and around the anterior face of the epiphyses of the surgical neck. Slight hyperostosis occurs in the posterior fossa of the left femoral head. Tibia and fibula are within normal limits. Both calcanea show slight overgrowth on the external surfaces but there is no true arthritis or osteitis.

This appears to be a case of limited mixed osteo-arthritis of the lower thoracic, the upper lumbar vertebrae and early changes in the femoral heads.

CERVIDAE

Red Deer (*Cervus elaphus*). ♂ P.Z.G. 12,125

Menagerie exhibition 86 months. Killed because of poor condition. The skull was too damaged at the time of death to be saved for examination. The cervical spine is within normal limits. The insertion of the longus colli on the 7th shows some elevations as if calcification had taken place at the root of the insertion. Thoracic 1 shows slight irregular downward lipping and the 2nd has a little superior roughening. This corresponds with the thoracic fibers of the longus colli.

The 6th and 7th thoracics show marked anterior lipping and there is a suggestion of close apposition of these two bones by reason of slight eburnation at these anterior overgrowths. The 7th and 8th show porotic destruction of their interfaces and an extension of the hypertrophic periostitis around the insertion of the rib at both sides. These two bodies had been coaptate as indicated by eburnation of part of the surface. These two joints therefore have moved but were the seat of ulcerative and hypertrophic arthritis. Between the 11th and 12th there is slight anterior lipping, more on the right side. Also slightly between the 12th and 13th. The lumbars, the sacrum and pelvis, the scapulae, breastbone and ribs, are negative.

The humerus and radial joints are negative except for some superficial erosion of the ulnar cartilages and the thin sheet of bone on the line of fusion of the radial segments. This is best shown on the right side. The posterior extremities are within normal limits.

It is interesting to observe that the somatic skeleton, humeri and femora are extremely light in weight and have a dull tone on striking together. The metatarsals and metacarpals are distinctly heavier with a sharper tone when struck together. The hind feet are within normal limits. The right forefoot has slight periosteal exostosis on the plantar and latero-plantar surfaces of the proximal phalanx. The middle phalanx has excessive periosteal overgrowth of chronic inflammatory type with porosis, wearing down and eburnation of the distal articulating surface. This is worse in the external middle phalanx than in the internal. The left foot has some periosteal overgrowth all around the body of the internal middle phalanx.

This animal therefore has slight hypertrophic spondylitis of the 1st and 2nd thoracics and 11th and 12th thoracics. There is ulcerative spondylitis of the 6th, 7th and 8th. Hypertrophic and eburnating arthritis of both anterior phalangeal joints.

Sambar Deer (*Rusa unicolor*).   ♀   P.Z.G. 12,092

In menagerie exhibition 164 months. Killed because of poor condition from old wound. The spinal column is negative except for erosive changes in the interfaces of the 12th to 14th thoracics. The anterior shoulder girdle, the anterior extremities, the pelvis, and the upper part of the femur are within normal limits.

The left patella shows periosteal overgrowth along the insertion of the external strands of the ligament inferiorly. The epiphysis of the femur is roughened by periosteal overgrowth along the insertion of the lateral ligaments externally, running up the insertion of the capsular imprint, anteriorly just above the articulating edge. The inter-condylar space is within normal limits but the margins of the condyles are roughened. The articular cartilage must have been removed from the internal condyle and the internal tuberosity of the ilium. There is no porosis but distinct eburnation. Luxation did not occur, but deformity almost certainly caused the foreleg to become twisted in order for the condyle to be accommodated on the surface of the tuberosity. The second metatarsal shows some surface erosion without hyperostosis. The bones below are apparently within normal limits. The right lower femoral end is distinctly worse but the lesions are comparable. This leg was probably fixed because an area of eburnation on the tibia is much smaller than on the left side. The lower epiphysis of the femur shows some porosis combined with hyperostosis. Overgrowth along the internal edge of the patella is more extensive on the right than on the left. All the metatarsals show erosion along the insertions of the ligaments but not on their articulating surfaces.   There has been dislocation or subluxation of some sort because eburnation exists in one of the cuneiforms.

This animal therefore had a chronic peri-arthritis and subluxation of both knees. Spinal changes are limited to erosive arthritis of the lower thoracics.

Sambar Deer (Rus unicolor).   ♂   P.Z.G. 11,878

Wild born male that had been on exhibition 237 months and was shot because of poor condition, found to be due to chronic tuberculosis of the abdominal tissues and arteriosclerosis.   There was roughness of the ventral surface of all thoracic vertebrae with lipping on the ventral surface from the 8th to 13th.   No ankylosis.   There was exostotic growth on the mammillary processes of the 1st, 2nd, 3rd and 4th lumbars, so involved that they cannot be separated.   Slight lipping on the ventral surface between 1st and 2nd and 3rd and 4th lumbar vertebrae.   The spinous processes are within normal limits throughout. The articulating surfaces of the ulnae show early ulceration to the spongy bone.   The epiphyses are within normal limits laterally.   The knee joints are negative as are the smaller joints.   The skeleton is malacic.

American Elk (Cervus canadensis).   ♂   P.Z.G. 3,878 ·

Captive bred male, on exhibition 91 months, that gave a history of losing flesh for some time and becoming stiff in the hind quarters.   It died finally from traumatism but was found to have a chronic nephritis, fatty infiltration of the liver and acute pulmonary arteritis.

Both humero-scapular joints, right upper and lower femoral joints, show chronic inflammation of the fimbriae and articular cartilages with edema in the peri-articular tissues. There was a chronic bursitis of the right knee.   The lesion in the right humerus was farther advanced so that necrosis had occurred in the epiphyses to a degree that caused separation of this layer of bone.

Mouflon (Ovis musimon).   ♂   P.Z.G. 2,246

Wild born male on exhibition 144 months.   Stiff and failing some time before death which resulted from chronic myocarditis, calcareocaseous tuberculosis of the right lung, chronic diffuse nephritis.   This was a chronic ulcerative process of both tibio-tarsal joints which showed disappearance of synovia, erosion of the plate of bone of the epiphyses and exposure of the spongy tissue.   The peri-epiphyseal tissue was not involved.

BOVIDAE

Goat Antelope.   N.M. 258,652.   (Fig. 1)

This specimen was sent to the National Museum by the Rev. Dr. D. C. Graham of West China Union University, Chengtu, Szechwan, China.   Capricornis sumatrensis milne-edwardii, female.   Identification by and bones lent by Dr. Gerrit S. Miller, Curator of Physical Anthropology, National Museum, Washington, D. C.   This is believed to be a truly wild animal and to have been shot in the wild.   The existence of this species in menageries is not known.   This skeleton offers evidence of chronic hypertrophic arthritis of lower cervical and upper thoracic vertebrae, the shoulder complexes and hip joint, the last resembling malum coxae senilis of man. This must have occurred in the wild, the subject being a running and jumping animal.

There is moderate lipping of the thoracic and lumbar vertebrae. The costal processes of the upper thoracic are hypertrophic and the corresponding point facet of the rib is surrounded by hyperostosis. Where the greatest degree of hyperostosis has occurred there is eburnation of the vertebral faces and the corresponding surfaces of the ribs. The cervicals are not involved.

Both glenoid cavities of the scapulae show eburnation and there is slight overgrowth around the lips. Both humeri show eburnation at the lower interior part of the head of the bone. The contours of the glenoid face of the humeral ball indicate an upward dislocation on both sides. The head of the humerus is, on both sides, surrounded by definite lipping, widest around the inferior half of the ball. Hyperostosis occurs across the main coracoid notch. The entire picture of the shoulder joints is that of hypertrophic osteoarthritis with subluxation that can be called ulcerative at no place. The lower joints of the anterior extremity are not involved.

Femur. The head on both sides is worn down by eburnation; on the right the cancellated interior is exposed. In addition, the wearing away has been so great that the insertion of the ligamentum rotundum has been reduced from a diameter of normal 12 mm. to 3 mm. It possibly did not exist as a strong tissue. The appearance of this femoral head is that it has been depressed to a right angle but this is due to the wearing away of the top of the head rather than to a change of true angulation of the head and neck with the longitudinal of the bone. The surgical neck and the epiphyses show hyperostosis. The acetabulum shows marked hypertrophic change with ulceration in the depth. The greatest degree of hyperostosis has taken place around the antero-inferior lip. It would be opposite the greatest degree of hyperostosis of the femoral head.

Nylghaie (Boselaphus tragocamelus). A.M. 5,526

In menagerie exhibition 101 months. Both posterior ankles show superficial ulceration in the middle of the tibial articulating surfaces and on the corresponding surfaces of the astragalus and scaphoid. No periarticular changes. Anterior extremities are without material change, slight roughening. This appears to be an early stage of infectious internal arthritis.

Leeche Antelope (Kobus leche). ♂ P.Z.G. 9,592. (Fig. 2)

Wild bred male on exhibition 110 months. Dying from acute gastritis and pancreatitis, soft stone in the right renal pelvis and ureter, chronic catarrhal cystitis and trichurus. There is a productive spondylitis with very great bony overgrowth from the 6th cervical to the 7th thoracic. The 4th, 5th and 6th thoracics are thoroughly ankylosed by calcification on their free ventral surface and of their mammillary processes and about the superior process of the lateral process which makes up the costal articulation. The articulating position of the 1st, 2nd and 3rd ribs has been surrounded by an enormous deposit of irregular bone so that the rib must have been loose. (Ribs discarded unintentionally.) This hollow is 4 cm. across and 2 cm. deep; in the first rib socket, surface is rough, not eburnated. This is a severe advanced case of calcifying ankylosing spondylitis of the thoracic type.

Radiograph 9,592 shows densities at all intervertebral articulations, and the lower cervical and first thoracic so thickened by overgrowth that the intervertebral line is not seen.   Marked hypertrophic growth on anterior surface of the bodies causing false ankylosis of the 4th, 5th, 6th and 7th thoracics.   Some destruction of the anterior half of the 4th thoracic vertebral body.   Posterior articulations also opaque.   The hypertrophy on bone about the costal insertions easily seen.

In contrast with the backbone of a deer believed to be normal grossly, there is irregularity of bone deposit in the diseased spine, also probably some demineralization of the main parts of the vertebral bodies.

Yak (*Poephagus grunniens*).   A.M.

Productive periosteitis of all parts of the vertebrae of the lower cervical region and perhaps two of the thoracic.   No ankylosis.   Transverse processes, posterior arch and spine involved in all cases.   Cordal canal not seriously involved.

Yak (*Poephagus grunniens*).   N.M. 14,328

Both femorotibial joints with wider tuberosity at lips than normal, probably internal subluxation of both tibiae.   Eburnated plaque on both tibial tuberosities.   Both patellae have marked roughening of the anterior surfaces, greater than that of other Artiodactyla.

Yak (*Poephagus grunniens*).   N.M. 174,734

All cervicals show some lipping of the inferior or ventral edge.   Thoracic shows marked overgrowth with articulating surface ulceration.   Metatarsals and metacarpals show hyperostotic, epiphyseal growth, also slight overgrowth on related carpals and tarsals.   No damage to the articulating surface proximal or distal.   All ribs show slight overgrowth at joints—not on articulating surfaces.   Phalanges show slight ossifying periosteitis but no true arthritis.

Yak (*Poephagus grunniens*).   A.N.S. 3,078.   (Fig. 3 and Pl. III)

Advanced osteosclerosis of all posterior joints including the pastern and down as far as the primary digit.   Some ulceration of every articular surface, in some places ¼ inch deep.   Anterior extremities much less involved.   First costo-sternal joint shows swelling with deep ulceration of both bones.   Along entire column; cervicals—no involvement of bodies; thoracic—involvement of bodies, lipping and with ankylosis.   Some articulating surface ulceration in lower cervicals.   Caudal only slightly involved, but caudal-sacral joint deeply ulcerated.   All tarsal and carpal bones hyperostotic.   Patella, surface hyperostosis without ulceration or facets.   Subluxation of knees but apparently no malposition of elbows or shoulders.
Beautiful case.   Mixed ulcerative and osteosclerotic arthritis.

Yak (*Poephagus grunniens*).   ♀   P.Z.G. 12,089

In menagerie exhibition 148 months.   Dead from senility.   The head and backbone are within normal limits down to the last lumbar which shows ulceration of its sacral face and eburnation on one side and apposed to it is the anterior face of the sacral in which a similar lesion exists, but not so far advanced.   There is perhaps some hyperostosis about both bones. There is therefore an ulcerative, eburnating spondylitis of the last lumbar and the first sacral.

The pelvis, ribs and scapulae are within normal limits. The extremities, except as follows, are within normal limits. The middle third of the articulating surfaces of both radii show some depression, particularly on the ulnar side, as if the cartilage and bony cortex immediately below had been soft and possibly had suffered early ulceration. The upper ends of both femora are within normal limits. The knee joints on both sides show periosteal thickening around the insertion of the joint capsule and edge of the articular cartilages. The joint surfaces are within normal limits. The distal articulating surfaces of both tibiae show superficial erosion at the middle area, comparable to the erosion in the ulnae. The other extremities are within normal limits.

Yak (*Bos grunniens*). A.M. 88,106

All but the middle lumbar involved with ulceration, eburnation; ankylosis of the upper lumbar and lowest two thoracics.

Indian Buffalo (*Bibos indicus*). N.M. 48,812

Ossification of the anterior ligament in the thorax with ankylosis of three vertebrae, probably 8th to 10th. Lumbar similar with marked hyperostosis all along the spine on the mammillary processes where there is the insertion of the extensor longus. None of the joints of the extremities shows any change on articulating surface. All bones are light as if atrophic.

Indian Buffalo (*Bubalis bubalis*). ♂ P.Z.G. 12,380

In menagerie exhibition 223 months. Killed because of poor condition. The skeleton in general is slightly demineralized, best shown in the skull and sternum. The skull is within normal limits, except that it is quite light. The teeth are quite loose but it is impossible to state that there had been necrosis at the roots.

The vertebrae are negative down to and including the 6th cervical. The 7th shows slight hyperostosis around the posterior articulations, especially the 1st thoracic. Beginning with the thoracic first and extending to the sacrum there is hyperostosis about the posterior articulations. This is worst in the 1st to 7th, inclusive, after which the amount of hyperostosis decreases. The articulations at the roots of the spines of the 1st and 2nd, 2nd and 3rd, 3rd and 4th, 4th and 5th, and 5th and 6th show roughening and also eburnation and irregularity of the articulating surfaces. There must have been an inferior subluxation at the joints of these vertebrae. The heads of the 4th and 5th ribs show eburnation and hyperostosis.

On the anterior surfaces of the bodies of the vertebrae conditions are negative until the 3rd thoracic is reached and then from the 3rd to the 7th there is superior and inferior lipping. The intervertebral surfaces are within normal limits. The 1st, 2nd and 3rd lumbars show slight ossification in the position of the anterior ligament. There is eburnation without porosis or hyperostosis at the lumbo-sacral joint.

The ribs, other than those mentioned, are within normal limits. There is slight hyperostosis on the posterior end of all the ribs corresponding to the insertion of the erector spinae group. The sternum is very light but seems not diseased.

The left scapulo-humeral joint must have been subluxated inward of the humerus since there is eburnation of the external edges of both glenoid surfaces of the head of the humerus. The elbow complex is within normal limits. The right shoulder complex must have been dislocated as on the left side since eburnation is seen in corresponding places.

The pelvis is negative except for hyperostosis about the supero-posterior angle on both sides at the beginning of the ilium. This corresponds with eburnation of the anterior fourth of the femoral head showing that there must have been a postero-superior subluxation on both sides. On the right side there is definite porosis of the head of the femur adjacent to the eburnation and on the unarticulated part but within the line of the joint capsule. This is slight on the left side and at corresponding places. The lower legs, fore and aft, including the digits are within normal limits.

We therefore have in this animal a hypertrophic and ulcerative spondylitis, principally of the posterior articulations, most marked from the 2nd to the 7th thoracic, decreasing markedly toward the sacrum; subluxation of both shoulder and hip complexes without great hypertrophy but with porosis of the articulating surfaces.

PINNIPEDIA

Sea Lion (*Zalophus californianus*). ♂ P.Z.G. 11,982

In menagerie exhibition 141 months. Death from chronic gastro-enteritis. Skull negative, except mandibulo-glenoid joints. Both sides, but chiefly the left, show erosive, porotic arthritis of the articulating surfaces. There is no hyperostosis, eburnation or subluxation. The spinal column, scapulae, fore extremities and ribs are all negative. Tarsi and carpi and meta-bones and phalanges appear to be within normal limits. The right pelvis and femur are within normal limits, with possibly one exception. Across the surgical neck from the epiphysis of the head to the top of the greater trochanter there is a slight lipping and roughening overgrowth. The left pelvis and femur are the site of arthritis, subluxation, hyperostosis and eburnation. The acetabulum is flattened and shallower than normal, chiefly towards the upper posterior portion which has been worn away and eburnated by subluxation. Each of the acetabular edges shows definite overgrowth. The corresponding portion of the head of the femur shows polished marble-smooth eburnation. There is an upper and posterior dislocation. There is no sign on the head of the left femur where the ligamentum rotundum was inserted; the location on the right side is very faint. The posterior face of the femoral head is irregularly covered with hyperostosis down to the inter-trochanteric line. The lower hind legs are within normal limits. Chronic hypertrophic arthritis, femoro-pelvic.

CARNIVORA

Ursidae

Bear (*Selanarctos thibetanus*). N.M. 240,668

Shot wild. Both humero-ulno-radial joints are involved by epiphyseal exostotic overgrowth without apparent abnormalities to the articulating surfaces. Both knee joints show similar overgrowth of bone but only on the left is there any obvious damage to the articulating surface. The internal condyle of the tibia shows some ulceration. The internal condyle of the femur shows a baring of the spongy bone. Both bones show near to the above defect a slight eburnation. There is growth and overlapping of the radial edge on both sides. Incomplete skeleton. Hypertrophic arthritis, elbow and knee complexes.

European Brown Bear (*Ursus arctos*). ♂ P.Z.G. 11,954. (Fig. 5)

Exhibition 288 months. Dead from broncho-pneumonia. Spine negative to the 3rd cervical, 3rd and 4th show marked lipping, 5th apron-like lip to 6th. Sixth, 7th and 1st thoracic show marked hyperostosis of the body, 2nd and 3rd slight lipping, 6th lateral apron-like lipping. From the 8th to 14th, inclusive, there is slight lateral lipping. The 14th and 1st lumbars are slightly wedge-shaped so that the bodies cannot be apposed. Definite lipping is seen in the remainder of the lumbars, worse in the 6th on the superior edge. Posterior surface—osteoporosis of the axis. Thoracic 1st, 2nd, 3rd and 4th show hyperostosis of the posterior arch and some lipping to posterior articulation. Thoracic 13th and 14th and all lumbars marked hyperostosis of apophyses and posterior articulation. Osteoporosis of the posterior articulation of lumbars from 1 to 5, inclusive, thoracic 11 to 14, inclusive. Osteoporosis in the bodies of the 7th cervical and 1st and 2nd thoracics.

This is a serious combined form of spondylitis in the lower cervical and first two thoracics, and from the 12th thoracic to the pelvis. The animal almost certainly had a hump in the back at mid-lumbar region because of the wedging of some number of vertebrae. The pelvis is completely ossified. There is no marked change in the sacrum and the coccyx is not involved.

The skull had been finely treated by rapidly moving steel but had been probably within normal limits. There are many worn teeth but the alveolar processes are probably within normal limits. Scapulae are within normal limits as are the ribs, except the top rib on both sides, which shows marked hyperostosis and porosis on the vertebral but not the anterior end. The left shoulder complex is negative. The elbow complex shows marked hyperostosis around the joint with ulceration of the joint surface. There is marked overgrowth of the coronoid process of the ulna. The lower end of the bone is much less involved. The carpal bones are not involved or at least only show hyperostosis on the scaphoid. The metacarpals show overgrowth of the epiphyses and the proximal end of the 1st phalanges is also involved. Articulating surfaces are within normal limits. The right shoulder is negative. The right elbow is involved comparably to the left but much less extensively. The right carpals and metacarpals are practically negative. The lower posterior extremity is within normal limits.

X-ray of cervical 7 to thoracic 5—opacity of the bodies of the first two bones, slight thickening and irregularity of the inter-vertebral lines of 2nd and 3rd thoracics; slight rarefaction of the bodies of the 3rd and 4th and possibly 5th. Lateral view shows general reduction in translucency of the entire 7 vertebrae. Intervertebral discs possibly not destroyed except between 2nd and 3rd thoracic. Mid-lumbar thickening at intervertebral lines, greater than normal translucency of main parts of the body. Marked marginal lipping on lumbars.

There appears to be in this bear an extensive productive and porotic arthritis affecting notably the spine but also the anterior extremities below the shoulder complex.

Sloth Bear (*Melursus ursinus*). N.M. 22,965

Rear knee epiphyseal osteoarthritis; slight lipping at the margin of articulating surface on both bones, hypertrophic change on the patellar articulation of the femur. Anterior overgrowth around edges of the patella. Articulating surfaces of patella and femur very rough, some of it probably artificial. Vertebrae: posterior edge of the ventral surface of the 7th cervical vertebra shows an irregular ulcerative defect of about 1.5 cm. in width, 5 mm. antero-posteriorly and the bone is eaten away 2–3 mm. The first lumbar also shows a little lipping. Just lateral to the mid-line of the ventral surface of the 3rd and 6th thoracic there is a nubble that grows backward like a lipping. Slight lipping is shown on the anterior tip on the articulating edge of the ulna on both sides. Hypertrophic arthritis —knee and vertebrae.

Black Bear (*Ursus americanus*). ♀ P.Z.G. 11,800. (Fig. 7)

In menagerie exhibition 167 months. Killed because of poor condition. Many pathological tissues found suggestive of chronic infection in a senile animal.

The first 5 thoracics show slight lipping about the insertion of the ribs and roughening about the mammary articulation of the ribs. Sixth, 7th and 8th vertebrae are completely ankylosed; 8th to 14th show distinct lipping, that on the 11th being an upward apron. There is wedging of the 13th and 14th bodies. There is a very slight turning out around the first 4 lumbars. The 6th lumbar and the 1st sacral show distinct lipping. The lateral aspect shows the greatest change around the insertion of the ribs, 6th, 7th, 8th and 14th. The posterior articulations of the 2nd, 3rd and 4th lumbars are involved in hyperostosis. The articulating surfaces of practically all posterior articulations are slightly porotic. This is least marked in the cervical area. The pelvis shows slight hyperostosis, anterior-inferior edge of both acetabula. Completely ossified pelvis. There is some lipping of the caudal vertebrae.

X-ray of 6th, 7th, 8th, 9–11th thoracics, lateral view. Opacity of the posterior arch around the costal insertions with absorption at the base of the costal insertion of the 6th and 5th. Intervertebral disc absent between the 6th and 7th, probably not destroyed between 5th and 6th, bones kept separate by ossification on the anterior face. Increased density of the bodies and posterior arches of 9th to 11th, with marked lipping over the superior intervertebral space, from the last numbered vertebrae.

The skull is badly damaged by shot. There is marked osteoporosis of both maxillae. The posterior extremity shows slight thickening around the insertions of all capsular ligaments. Marked roughenings at the epiphyseal line, both upper and lower extremity. The left fibula is attached at its upper end to the external condyle of the tibia. The articulating surfaces are within normal limits, as are also the humeri. The right radius and ulna combination is practically normal. The left is negative at the upper end. The lower end shows overgrowth within the capsular zone and about the epiphyses. The right carpus shows overgrowth, particularly of the scaphoid and cuneiform group. The metacarpals show marked hyperostosis at the insertions of the tendons, that is, along the sides of the shafts and a little at the epiphyses, which is sharply marked on the 1st, 2nd and 3rd metacarpals. The 1st, 2nd and 3rd phalanges of the first four toes show hyperostosis at the insertion of the tendon along the lateral edges on the palmar surfaces and slight overgrowth on the epiphyseal margin. The right carpus is similarly but not so much involved. The right metacarpal bones 1 and 2 show hyperostosis at the base and within the capsule, but little where the tendons are inserted. The phalanges are irregularly involved, No. 2 on the right showing some lipping superiorly. The second row of phalanges are involved on both sides.

The tarsal bones are almost within normal limits, there being slight roughening in the calcaneal point. The metatarsals show some roughening of the proximal part of the shafts. The second and third rows are almost within normal limits. This is apparently a hyperostotic periostitis more than arthritis.

We have therefore a case of chronic, hypertrophic osteoarthritis involving the vertebral joints and the shafts and the epiphyseal arches of the hands and feet.

Brown Bear (*Ursus arctos*). A.N.S. 6,264

The ventral surfaces of the vertebrae, cervical and lower thoracic, including the first dorsal and sacral joint, show hypertrophic periosteitis with slight roughening of the articulating surfaces but without definite overgrowth. Dental caries and alveolar porosis.

Alaskan Bear (*Ursus gyas*).   ♀ P.Z.G. 11,789. (Plate III)

In menagerie exhibition 385 months. Senile, killed. Wild born female. Poor appearance for some months. No knowledge of arthritis. There was a tumor at the tail of the pancreas, cysts of the thyroid and a moderate grade of medial arteriosclerosis. The vertebrae are involved from the axis to the pelvis. In the cervical region there is marked overgrowth of the ventral surface that seems to stream along as if following the anterior ligament. The 5th, 6th and 7th are nearly ankylosed by this lipping. The thoracic vertebrae show also irregular lipping with some overgrowth around the costal joint. The 11th and 12th thoracics are ankylosed by this overgrowth. The lipping is lateral to the midline in the main chest. The lumbars show marked lipping, being worse in the 4th, 5th and 6th where they are ankylosed. The 1st, 2nd and 3rd ribs show slight mushrooming at their articulations. The scapulae are negative. The costal articulating surfaces of 3rd to 5th vertebrae show hyperostotic overgrowth. The intervertebral surfaces are not involved. Extremities are negative.

Hypertrophic spondylitis entire spine.

Grizzly Bear (*Ursus horribilis*).   ♀   P.Z.G. 11,990

In menagerie exhibition 296 months. Mammary carcinoma. The vertebral column is negative to the 6th cervical. Lipping on the 7th, superior and inferior. The 1st thoracic shows considerable superior lipping, being very irregular with granular hyperostosis. From the 7th to the 10th there is flaring of the discal margins. The 7th thoracic shows lipping in the form of an apron on the right side covering the place of the disc. The 11th to the 14th thoracic show marked hyperostotic lipping. The 1st lumbar is least marked of the vertebrae. The 2nd to 6th lumbars inclusive show marked granular ridging. The posterior arches of spine and ribs are within normal limits. The pelvis is negative, apparently that of an old mature animal. The skull is negative. The scapulae are remarkably light in weight but show no evidence of arthritis. The humeri are within normal limits.

The right ulnar and radial shafts are within normal limits to the last one and a half

inches.   Both bones show periarticular and hyperostotic roughening about the epiphyses apparently within the capsular ligaments.   The articulating surfaces are within normal limits.   Lesions are greater on the flexor surfaces.   The carpals all show periostotic changes, but none on the articulating surfaces.   The meta-carpals 1–4 show marked hyperostosis around the proximal joint so that there must have been a destruction of the interosseous ligament, at least between the 1st and 2nd, and the 2nd and 3rd.   The lesion on the palmar surface of the 1st metacarpal is the greatest.   The left radius and ulna are negative down to the epiphysis where the capsular ligament was attached.   The carpals on the left side show low-grade hyperostotic growth, particularly on what appears to be the external cuneiform.   The left side is not so advanced as the right side, not that the lesions are not present, but that they are not nearly so pronounced as on the left side.   The tarsals are not available.   The metatarsals are slightly hyperostotic but not by any means so definitely as are the metacarpals.   The interosseous membrane of the 3rd and 4th and the 4th and 5th was ossified.   The plantar surfaces of the first and second right and all five left show some moderate hyperostosis.
General hypertrophic arthritis.

HYAENIDAE

Hyena (*Hyaena hyaena*).   ♀   P.Z.G. 12,109

In menagerie exhibition 182 months.   Thyroid disease, gangrene of foot.   Killed.

Atlas and axis are within normal limits.   Beginning with the 3rd cervical, the ventral surfaces are rough and show slight porosis.   The posterior articulations are also rough as if the synovial membrane had been worn through.   The 6th and 7th cervicals show great porosity of the ventral surfaces and some slight lipping of the 6th.   The thoracic vertebrae show slight lipping and very great porosity of the ventral surfaces.   The posterior articulations are slightly roughened and porotic in the upper six costal joints and are worn as if the synovia had been damaged.   The costal lateral processes are porotic even though, as in the lower thoracic, the joint faces of the body are smooth.   The main spines are all porotic and roughened and from the 2nd to the 8th are distinctly hyperostotic.   The lower thoracics remain rough but become most so at the 14th where uniform lipping becomes marked with bony deposits between the 15th thoracic and 1st lumbar.   These two vertebrae are wedged-shaped.   There is porosis of the inter-vertebral faces.   This continues and becomes worst in the 2nd and 3rd lumbars.   The 3rd is very badly diseased.   The 5th is within normal limits as are the pelvis, sacrum and tail.   The 6th caudal is irregularly and slightly porotic as if it had suffered an injury.   The sternum is as light as a piece of common blotting paper.   The posterior ends of all the ribs are roughened and those joining with the upper thoracic are distinctly porotic.   The scapulae are within normal limits.   The humeri are within normal limits at the upper ends and the two elbow joints may be described together as showing marked periarticular hyperostosis with porosis with lipping from the articulating edge.   On the left side there is distinct internal eburnation and on the right side a beginning change in the same location.

The metacarpals and metatarsals appear to be within normal limits.   Many phalanges are missing.   The posterior first metatarsals can be identified and one shows marked periostosis and eburnation of the proximal end.   The two second phalanges, probably posterior, show marked periostosis without change in the joint surfaces. These are the only observations upon the toes of this animal that can be made, as he had chewed off the remainder from the hind feet. It is possible and indeed probable that he chewed them because of parethesia from the arthritis. The leg and fore-leg bones proper are within normal limits.

This animal had marked demineralization and mixed arthritis of the whole spine with principal locations in the lower cervical, the lower thoracic and mid-lumbar regions. Also osteitis and probably arthritis in the small joints at the end of the posterior extremities. It appears to be a mixed hypertrophic and ulcerative arthritis.

Hyena (*Hyaena sp.*).   N.M. 155,455

Very slight lipping of the lower thoracic vertebrae. Both scapulo-humeral joints involved. Both articular surfaces are ulcerated and show cancellous bone. There is slight lipping around the glenoid cavity. Early ulcerative arthritis.

Hyena (*Hyaena sp.*).   N.M. 172,685

Slight roughening of the 6th cervical ventral surface. The middle thoracic (numbers not clear) shows marked lipping with erosion and eburnation of the opposed bodies. Mid-lumbar shows very marked overgrowth with porosity of the internal vertebral junctions. There is slight overgrowth between the left internal condyle of the humerus and the anterior promontory of the ulna. Marked osteoarthritis is shown in the lower end of both femurs with lipping at the edge of the popliteal groove. No frank dental caries is found, but there is absorption of the alveolar process.

Hyena (*Hyaena sp.*).   A.N.S. 11,962

In the lower thoracic and upper lumbar there is moderate ventral osteosclerosis with ankylosis of two pairs that appear to be mid-dorsal. The cordal tube is not encroached upon.

Hyena (*Hyaena sp.*).   A.N.S. 4,260.   (Plate III)

From the 3rd cervical to the sacrum there is productive osteitis confined very largely to the ventral surface and some to the transverse processes. The spines of the middle thoracics are joined together. The atlas, axis and sacrum are not involved. Only at mid-thoracic are the articulating surfaces ulcerated through to spongy bone. Ribs are within normal limits. Both scapulae and humeri show periarticular osteosclerosis and involve the outer end of the clavicle. There is eburnation and ridging of the shoulder joint. The right elbow joint is uninvolved. The left shows slight osteosclerosis of the internal condyle.

Hypertrophic arthritis with ulceration.

Hyena (*Hyaena sp.*).   A.N.S. 4,261.   (Plate III)

The posterior axis and anterior surfaces of the atlas, the middle cervicals, first two thoracics and one lumbar pair show ventral hyperostosis, the rest of the bone being within normal limits. The articulating surfaces are within normal limits except at one thoracic joint where there has been complete ulceration and exposure of the spongy bone.

Hypertrophic arthritis with ulceration.

Hyena (*Hyaena sp.*). A.N.S. 11,963

The lumbar vertebrae (all bones are not present) show hypertrophic changes on the ventral surfaces and also on the transverse processes laterally and posteriorly. There was probably no damage to the articulating surfaces of the body. The cordal canal was not interfered with. The thoracic and cervical vertebrae and skull are within normal limits.

Hyena (*Hyaena sp.*). A.N.S. 12,021

Incomplete skeleton. The posterior level of the right femur is involved, as are both anterior extremities, right humero-ulnar joint, in osteoporosis and hypertrophy. There is slight osteoporosis of the outer end of the clavicle.

Hyena (*Hyaena sp.*). W.I. 7,102. (Figs. 8 and 9)

The pelvis is lacking. There is osteoarthritis of the anterior shoulder girdle and the anterior extremities. It is principally the exostotic type with some eburnation, especially remarkable grooving of the ulna where it articulates with the internal condyle. A wonderful example of productive and degenerative arthritis. There is no material damage to the long bones of the hind extremities. There is some periosteal bone formation and osteoporosis of the ribs. Thickening of external tubercle of the humerus, irregular marginal thickening and bony rarefaction of the internal condyle.

X-ray shows thickening and suggestion of opacity on the edge of the radius and ulna, probably both overgrowth and ulceration. Some demineralization of the wall with thinning of cortex, most marked near the upper head of the humerus.

VIVERRIDAE

Binturong (*Arctitis binturong*). ♂ P.Z.G. 12,159

The bodies of the lumbar vertebrae show hypertrophic osteitis with lipping but probably not complete ankylosis. The transverse processes are about normal and there are no evidences of damage to the articulating surfaces. The cordal canal has not been encroached upon.

Binturong (*Arctitis binturong*). ♀ P.Z.G. 11,821

Exhibition period 154 months. The skeleton is negative except as described. The bones are light but not particularly porotic. The cervicals are negative. There are 13 thoracics, and beginning with the 9th there is inferior lipping, irregular hyperostosis appearing on the ventral surface in the form of lipping and irregularities, to and including the first dorsal. The 13th thoracic and 1st dorsal were possibly ankylosed, and broken artificially since there is a layer of calcified bone on the dorsal that seems to be broken from the body of the thoracic. The remainder of the dorsals are within normal limits.

The left os innominatum is ankylosed along its anterior edge to the lateral processes of the 1st and 2nd sacral vertebrae. The posterior two thirds are free. The right os innominatum may have been similarly affected but there was a line of cleavage which permitted it to come apart in the softening process. The posterior parts of the vertebrae are within normal limits.

This is a hyperostotic and ulcerative spondylitis of the lower thoracic and 1st dorsal vertebrae with deep ossification of the bodies of the 13th thoracic and 1st dorsal.

FELIDAE

Jaguar (Felis onca). N.M. 12,296

Lipping and slight calcification of the lower thoracic along ligament with the upper dorsal.   Right humero-ulnar joint anterior process of ulna—erosion; external humeral condyle—slight erosion, both with ribbing like a screw and eburnation.

Jaguar (Felis onca). P.Z.G. 11,949

In menagerie exhibition 186 months.   Senility, abscess of lung and brain.   The skull is negative and the mandibles are within normal limits.   There is slight lateral lipping of the 2nd, 3rd and 4th thoracic vertebrae.   The general construction of the vertebrae is good.   Posteriorly there is no change.   The three-piece sacrum is within normal limits. The right humerus is negative down to the olecranon fossa.   The posterior edge shows slight roughening at the epiphyseal line.   The ulna and radius are within normal limits. The right metacarpus, carpus and phalanges are within normal limits, which is likewise true of the left.   The left humerus is negative down to the elbow which shows a definite periarticular overgrowth in the fossa.   The ulnar articulating surfaces are eburnated and the surface is worn off so as to reveal the cancellated bone.   The apposed articulating surface of the humerus has been worn to the depth of one eighth of an inch, more deeply on the right side.   There seems to have been no incorrect position of this ulnar joint and the rotation surface of the radius is not materially changed.   There is slight eburnation of the antero-inferior articulating surface of the radius.   In both legs the femur, tibia and fibula appear to be within normal limits.   The tarsus, metatarsus and phalanges seem within normal limits, except for the first proximal middle phalanx which is roughened at both extremities.

This seems to be a chronic hypertrophic osteoarthritis of the left elbow, one phalanx and very early lesions of the 2nd, 3rd and 4th thoracic vertebrae.

Leopard (Felis pardus).   ♂   P.Z.G. 11,910

Wild born male on exhibition 178 months.   It was crippled for many months although its appearance was good.   It was found to have chronic endo- and pericarditis, round ulcer of the stomach, chronic interstitial thyroiditis and calcification of one aortic valve.   Both radio-carpal joints and adjacent tendons showed calcareous deposits including the bursae. The articulating surfaces are dry and rough though not frankly ulcerated.   There were some bursa-like bodies containing clear gelatinous material and tiny calcareous granules.

Radiograph of jaws shows many broken teeth but little evidence of apical absorption or alveolar necrosis.   Upper canines are worst.

Leopard (Felis pardus). W.I. 3,681. (Fig. 11, a–c)

Elbow joint, all thoracic vertebrae, slight involvement above knee joint, all with hypertrophic osteo-arthritis.   Marked hyperostoses of 2nd and 3rd, 4th and 5th, 5th and 6th lumbars.   Roughening and irregularity of the intervertebral faces between 2nd and 3rd, 4th and 5th, 5th and 6th.   Hyperostoses of the inferior posterior articulations of 5th and 6th.

X-rays of lumbars show structure of bodies to be generally within normal limits. Slight thickening at the intervertebral ends of 2nd and 3rd, 5th and 6th. Definite hyperostoses at all joints that correspond to overgrowths on the black and white photograph. Marked irregularities of the intervertebral faces, 4th and 5th, upper face of the 6th and junction of the 5th and 6th.

Elbow radiograph does not show changes so well as photograph. Thickening around edges of the trochlea and adjacent fossae, thickening around the articulating head and coronoid notch of the ulna. Possibly slight decalcification of the subarticular ends of the same bones including the adjacent shaft.

This appears to be therefore a hypertrophic or hyperostotic and ulcerative lumbar spondylitis and arthritis of elbow.

Tiger (Felis tigris). A.N.S. 12,139

Productive osteoarthritis of the left elbow without apparent destruction of the articulating surfaces, some on the ventral bodies of the thoracic vertebrae. The cordal canal has not been encroached upon.

Tiger (Felis tigris). N.M. 22,258

From the 6th to the 10th thoracic inclusive there is marked lipping and it is worth noting that the lips are curved out with a space between them and the tip of the body of the vertebra. The intervertebral faces between the 5th and 6th, 6th and 7th, 7th and 8th, 8th and 9th and 9th and 10th are smooth, shining and eburnated. The spongy bone is exposed, the cartilage must have disappeared. The animal must have been humpbacked. The 4th, 5th and 6th lumbars show slight calcification at the place of the anterior ligament. This is an incomplete skeleton—other bones are negative. Mixed hypertrophic and ulcerative spondylitis.

Tiger (Felis tigris). ♂ P.Z.G. 12,101

In menagerie exhibition 222 months. Senility. The skull is within normal limits except for slight precanine maxillary absorption. Teeth are within normal limits.

There is slight lipping of the 1st and 2nd thoracic vertebrae on the upper edge of the bodies. Negative to the 5th lumbar, the edge of which shows slight lateral lipping on its sacral articulation. The sacrum and tail are within normal limits, as is also the pelvis. The ribs are all light in weight as judged by hand testing. Sawing through reveals a fairly uniform cortex slightly narrower than normal. The spongy structure is more extensive than usually seen in the ribs of carnivores.

The right scapula is light but the left is lighter. The right scapular head shows exostosis with necrosis around the infero-anterior edge of the glenoid. The articulating surface is worn away by porosis from a downward subluxation. The upper third of the glenoid surface almost certainly did not articulate with the humerus. The head of the humerus shows an eburnated inferior and posterior face occupying three fifths of the head. Around the articulating surface and following the tendon of the biceps there is osteophytic growth. The left scapular head is practically within normal limits, although there must have been some upward dislocation since there is an eburnated part of the humeral articulating surface at its extreme postero-inferior one-eighth. The remainder of the bones are within normal limits.

This tiger therefore has subluxation, hypertrophic and necrotizing arthritis of the right scapulo-humeral joint, very slight hyperostosis of thoracic and sacral vertebral bodies.

Lion (*Felis leo*). W.I. 7,100

Hypertrophic periosteitis of a few of the middle thoracic vertebrae on the anterior surface; no indication of ankylosis. Ankylosis of a vertebra near the end of the tail. Skeleton incomplete.

Lion (*Felis leo*). N.M. 12,319

Slight lipping of probably 5th and 6th thoracic vertebrae. Hypertrophic osteo-arthritis of humero-ulnar joint with eburnation and marked wearing-away of the articulating surface of the internal condyles. Marked grooving. Both elbows involved. This is an advanced case. Tarsals, etc., are negative.

PRIMATES

LORISIDAE

Peridicticus Potto. ♂ A.M. 52,698

Distinct lipping between 8th lumbar and 1st sacral. Intervertebral disc must have disappeared and partial ankylosis taken place. Otherwise the skeleton is perfect.

CERCOPITHECIDAE

Guinea Baboon (*Papio papio*). ♂ P.Z.G. 12,105. (Fig. 12)

In menagerie exhibition 184 months. Myocarditis, bronchiectasis, nephritis. Fully developed animal, male, with complete ossifications. Both glenoid cavities are roughened, the left one showing also hyperostoses on the malar projection. The maxilla shows four remaining teeth on both sides, two molars and two premolars. The anterior teeth are missing and there is irregular porosis and hypertrophy around the insertion of the upper left canine. Simple porosis right upper canine. The mandible has on the right one molar and three premolars. The 1st molar position shows porotic obliteration. The 1st premolar is distorted apparently by rubbing with the upper canine. Incisors are absent. There is no particular change where the sockets have been obliterated.

X-ray. The right upper jaw shows peridental thickening along the roots of the four remaining teeth. The left upper is almost identical. An old canine that has been broken off at the alveolar edge is visible above the molars in its usual position, somewhat absorbed along the edges and with an irregular canal. There is an irregular opaque mass on the alveolo-palatal border, nearer the mid-line on the right than on the left, that appears like an unerupted and somewhat absorbed molar. The positions of the incisors show several spots of opacity that are interpreted as scleroses resulting from an old infection. The right lower shows peridental absorption, as does the left, with a very moderate grade of thickening or sclerosis about the molars and one root of the first premolar. The position of the incisors and canines shows the position of the canine root without material change and porosity of the incisor.

Atlas. Lower articulating surface on the right slightly eroded at its extreme outer margin. The upper left articulation of the axis is lipped but not eroded. Slight lipping, lower edge of body of axis.

The cervical stretch has a slight convexity to the right including the 6th and 7th cervical and 1st thoracic. Slight convexity to the left including the 2nd, 3rd and 4th thoracic. Slight convexity to the right including the 6th, 7th, 8th, 9th and 10th where the convexity turns to the left including the 11th, 12th, 13th and 1st lumbar and then the 2nd, 3rd, and 4th lumbars bulge to the right. ·

The vertebrae were kept together so that the interfaces are not clear, but there is a sharp wedging of the bones that form the middle of each of the above curvatures. There is roughening along the position of the anterior ligament and there is slight lipping on the lateral faces of the vertebral bodies, most easily seen where the convexity and concavity are plainest.

The posterior articulations are everywhere roughened, worse between the 4th and 5th, 5th and 6th, 6th and 7th cervicals, 7th and 1st, 1st and 2nd and 2nd and 3rd thoracics.

This is low grade deformative spondylitis with complemental scoliosis resulting in a comparatively straight backbone. The corresponding ribs seem not to be materially changed in curvature. The posterior articulations of the 1st and 2nd ribs are slightly hypertrophic, whereas those of the 10th and 11th ribs are both hypertrophic and porotic. The 4th caudal vertebra is distinctly porotic. Some of the anterior spines are ossified tight while others have become loosened in the cleaning. The pelvis is not far from normal. The sacrum is completely ossified to the ilium. The anterior ends of the clavicles are slightly porotic and swollen.

The left scapula shows osteoporosis of the glenoid cavity and a complete ring of osteitis at the edge. There is overgrowth of the greater humeral tuberosity. The superior angle is abnormal but the degree to which the angle is normal of the humerus is not known to the observer. However, this superior bend is the same on the two sides. The trochlear surface is slightly eroded on its anterior face and there is hypertrophic change on its internal end. The olecranon is hypertrophic and porotic. The ulna is within normal limits. The radius has slight hyperostosis on its external face. Left Hand: All the four distal phalanges show osteoporosis and slight hyperostosis in their shafts and tips.

The conditions of the right scapula, humerus, ulna and radius are comparable to left. The right hand shows damage to only one distal phalanx. The small digit on the right foot shows hyperostotic growth in the two distal phalanges.

Femur. Both greater and lesser trochanter are higher than normal from hyperostotic overgrowth. The insertions of the lesser gluteus are ridged. The lower ends are rough but not distinctly pathological. Both patellae are hyperostotic on the anterior surfaces and lateral edges. The articulating surfaces are within normal limits. The tibial articulating surfaces are also within normal limits. The edges are slightly roughened. The fibula is ossified posteriorly at the superior end and completely ossified at the external malleolus. The articulating surfaces of the ankle are within normal limits.

There existed in this case, therefore, missing anterior denture, low grade, chronic, destructive and hypertrophic arthritis of the jaw, ulcerative and hypertrophic spondylitis with scoliosis, chronic hypertrophic arthritis of the scapulo-humeral and humero-radio-ulnar units. Hypertrophic osteitis of the patella and minor hypertrophic osteitis of the terminal phalanges.

Mandrill (*Mandrillus sphinx*). ♂ P.Z.G. 12,147. (Figs. 13, *a–e*)

In menagerie exhibition 154 months. Strangulated abdominal hernia.

Skull. Glenoid cavity very rough, the cartilage probably having been destroyed. There is hyperostosis extending to the mastoid. The mandibular condyles are very rough and with slight hyperostosis around them. The canine ridges are very rough and appear to be hyperostotic over the edge of the orbit to the front line and laterally down to the molars. There is also some roughening of the premaxillary and intermaxillary bone in front of the palate. Rough hyperostotic periosteal overgrowth on mandible below and around symphysis. The upper dentition consists of sixteen—2 molars, 3 premolars, 1 canine and 2 incisors. The denture of the lower jaw is similar.

X-ray. The left upper shows absorption at the root of the canine and all premolars with vertical peridental absorption. The right is similar. The plates of the canines are not clear as to their right maxillary position but it is probable that all four canines show some absorption along their edges. The left second incisor shows thickening and absorption vertically. The mandible shows slight vertical absorption of the first premolar on the left. The right is not clear. Incisors on right—absorption at roots of central both sides, possibly thickening of lateral incisor.

Atlas. Left anterior articulation shows periostitis on external surface. The internal surface is within normal limits. *Axis.* This is within normal limits. The 3rd, 4th and 5th cervicals show posterior hyperostosis and porosis on bodies. The 5th and 6th are ankylosed by complete junction of their bodies. Fig. 13ᴇ. The lower face of the 6th is porotic and there is lipping to the porotic and crushed 7th, which shows considerable destruction with hyperostosis. The 1st dorsal is likewise crushed, hyperostotic and porotic. There is a slight convexity to the left in the cervical vertebrae and a deformity, to the right, of the 1st thoracic. The 2nd and 3rd thoracics are ankylosed by marginal hyperostosis. The 3rd thoracic is wedge-shaped to the right as is the 4th so that there is a slight convexity to the left. The 6th, 7th and 8th are wedge-shaped to the left so that there is a slight convexity to the right. The 9th is wedge-shaped to the right which begins a slight curvature to the left. The 10th, 11th, 12th and 1st lumbar are ankylosed along the anterior ligament by bilateral overgrowth. There is lipping and porosity of the 2nd and 3rd junction. The 4th, 5th and 6th lumbars are ankylosed. There is interlocking lipping of the last lumbar with the 1st sacral. There is slight lateral lipping of the 1st and 2nd caudals. The left sacrum is ankylosed with the left ilium. The right is free.

The scapulae are within normal limits. The right arm shows slight hyperostosis about the internal edge of the semi-lunar notch. The coronoid end of the semi-lunar notch and the internal third of the trochlea are eburnated in such a way as to suggest a slight posterior subluxation of the ulna. No porosis.

Therefore we have a low-grade hypertrophic arthritis of both elbows with subluxation of the right. A similar condition exists on the left elbow where, however, the hyperostoses are greater and there is eburnation of the internal edge of the trochlear surface with the hyperostotic edge of the articulating ulna.

The pelvis and femur are within normal limits. The fibula on the right is united at its upper end by a complete ossification with the tuberosity of the tibia. The fibula on the left was in process of ossification but separated at the time of the preparation of the bones.

X-ray of Spine. Cervicals—irregularities of bodies of all 7 cervicals, especially marked in the 7th. Great irregularity of thoracic 1st, present also but less in thoracic 2nd, shadow between cervical 5th and 6th as if from false ankylosis, posterior arch thickening, posterior articulations obscure and irregular. Straight picture shows wedging of 1st, 2nd, 3rd and 5th thoracics. Hypertrophic overgrowth especially anteriorly with abolition of inter-vertebral line between thoracics 9th and 10th, 12th and 13th. Hypertrophic overgrowth between thoracic 13th and lumbar 1st, lumbars 1st and 2nd.

The distortion of the vertebral bodies, the roughening of their interfaces as described grossly, the roughened intervertebral lines as indicated by x-ray, all suggest combined hypertrophic and ulcerative spondylitis. The posterior articulations are involved in the cervical section.

We have, therefore, complete spondylitis including practically every joint in the column especially prominent on the anterior parts of the vertebral bodies. The posterior articulations are definitely but less prominently involved. There is extensive scoliosis with six definite curvatures. Ulcerative and hypertrophic spondylitis. Hypertrophic arthritis of the elbow complexes.

Papio Hamadryas.   ♀   Td. 1,342-B.   Wild

Skeleton negative except as follows. The anterior half of the 3rd cervical body had apparently no cartilaginous buffer because of the irregular porosity of its inferior face. Anterior digitoid lipping overhanging the superior anterior edge of the 4th. This vertebra also had irregular porosity of its upper face but less than the apposed vertebra. Cal-careous pieces, roughly 1 × 2 mm., lay between the digitations of the 3rd and the roughened anterior edge of the 4th vertebra.

Ulcerative and hypertrophic arthritis of the 3rd and 4th cervical vertebrae with prob-able disappearance of the intervertebral disc.

Orang Utan.   B-1,167.   Td.   Wild

Cervical vertebrae show some periostotic overgrowth of apophyses 6th and 7th. Porosis of the posterior articulation also. Lipping of the superior edge of the body of the 7th. Thoracic vertebrae show lipping of the superior edge of the 10th and 11th. Some wedging of these bodies. Skeleton otherwise negative.

Orang Utan.   ♂   B-623.   Td.   Wild

Odontoid process of axis and corresponding articulating facet in the internal anterior body of the atlas eburnated and ridged. No sign of porosis or necrosis. Lateral atlo-axial facets rough but not deeply porous. Anterior surface of both bones rough. Skeleton otherwise negative. This animal certainly had a stiff neck and perhaps it creaked.

Gorilla (*Gorilla gorilla*).   N.M. 23,373

The left knee joint shows definite hyperostotic lipping of the internal tuberosity and condyle, also of the periarticular edge of the femur at the patellar area. The largest point of the lipping is at the external edge of this junction of the patella and femur. The tibial tuberosity and the fibular junction are not materially changed. The articulating surface between external condyle and tuberosity shows marked hyperostotic area on the femur and roughening on the tibia where there was almost certainly damage to the cartilage. The internal condyle of the femur posteriorly shows disappearance of the bony plate, roughening of the surface, exposure of cancellated bone and beginning eburnation. The internal tuberosity shows on the articulating surface a slight exostosis and there is roughening on the opposed internal condyle. The external edge of the patella has a sharp lipping. The articular surface is rough and the cartilage was almost certainly worn away.

Chronic, mixed arthritis, monarticular.

Gorilla (Gorilla gorilla). W.I. 4,491

Hypertrophic periosteitis of right side of sacrum front and back. First and last foramina of sacrum obstructed. Hypertrophic periosteitis at junction of body and posterior ramus of the pubis on the right side. Left side similar with some irregularity of the body of the ischium.

Gorilla (Gorilla gorilla). ♀ B-1,756. Td. Wild

All epiphyses united. Probably corresponds with 28 human years. Vertebrae: 5th cervical, upper anterior articulating surface one third worn down, exposing cancellous interior; posterior two thirds within normal limits; posterior lower surface within normal limits, despite the fact that the anterior one third of the articulating surface of the 6th is worn away, and the posterior two thirds are as in the 5th. The inferior surface of the 6th shows some erosion of anterior one third but posterior two thirds within normal limits. There is slight lipping on the 6th but not on the 5th. The anterior half of the intervertebral disc was almost certainly damaged. This cervical group was probably bowed.

Gorilla (Gorilla gorilla). ♀ B-1,856. Wild

No record, but well matured specimen.

Third and 4th cervicals. Inferior surface of the 3rd cervical shows porosis and lipping extending out to left posterior articulation, which is slightly eburnated. The right side is within normal limits. The superior surface on the 4th cervical is almost exactly the same, but worse on the left side.

Second and 3rd Lumbars. These may have been porotic on respectively inferior and superior surfaces, but the skeleton is worn and therefore this is indefinite. This gorilla surely had a cervical pain.

Gorilla (Gorilla gorilla). ♀ B-1,851. Td. Wild

Young adult. Slight alveolar absorption. Teeth within normal limits. Slight overhanging lipping 9th to 12th thoracic vertebrae, inclusive. Some flattening of all and perhaps a slight wedging of 9th, 10th and 11th.

Right humerus, superior end within normal limits. The inferior end, anterior cavity, irregular ossified overgrowth fills it up. Posterior cavity, ossified overgrowth in depths. The edge of the anterior surface is osteophytic. Ulna. The posterior coronoid is rough and lipped. The anterior coronoid is rough and overgrown and the anterior surface is porotic. Both radii are within normal limits. Left humerus, superior end, is within normal limits. Inferior end, anterior and posterior cavities are filled as on the right side. The anterior surface is porotic and grooved and the median ridge is porotic over a short narrow line.   There is an osteophyte on the ulna, anterior surface, which corresponds to porosis on ridge and to grooves on radial side.

Gorilla (*Gorilla gorilla*).   ♀   B-1,854.   Td.   Wild

Old animal.   Cervicals incomplete.   Third and 6th show marked superior and inferior lipping of dentoid form.   Discs were possibly within normal limits.   Ninth to 11th thoracics show slight but definite superior lipping, also slight superior on 13th.   The tibio-fibular joints are ossified but are not continuous or ankylosed, since they separate without breaking.   There is excessive lipping in the cervicals, but the whole story is not available.   The thoracic lesions are not definite.

Gorilla (*Gorilla gorilla*).   ♀   B-1,798.   Td.   Wild

Compared with man about 30 years old.   Slight but clear superior anterior edge lipping of 9th to 13th thoracics and of inferior edge of 10th.   Inferior edge of 3rd and superior edge of 4th lumbars slight but clear lipping.   Separate spicule on left side of 4th on superior edge.

Gorilla (*Gorilla gorilla*).   ♂   B-1,954.   Td.   Wild.   Old animal.   (Figs. 15, a–e)

The 1st and 2nd lumbar vertebrae show abnormal lipping without wedging of bodies. There is slight but definite porosity on the posterior half of the intervertebral faces.   The 4th lumbar is ankylosed to the sacrum by right lateral ossification, in the position of the lateral ligament.

The sternum is within normal limits but the sternal end of the 1st rib is swollen and almost like a cauliflower, necrotic and brittle.   The left elbow is porous in the internal face of the humero-ulnar articulation.   Hyperostotic growth on the articular face of the ulnar concavity corresponds to the porosity on the humerus.   The right elbow and left carpus are within normal limits.

The left metacarpus—distal ends of the 1st, 2nd, 3rd and 4th show irregular swellings certainly from periarticular hyperostosis and a porosity probably from deep demineralization.   The 2nd and 3rd metacarpo-phalangeal joints show subluxation with eburnation. The 1st, 2nd, 3rd and 4th phalanges show proximal exostosis, the 2nd with porosity.   The right hand conditions are almost identical, excepting location.   The 1st, 2nd, 3rd and 4th metacarpal distal ends and the 1st and 4th show subluxation.   The 1st phalanx and the 1st, 2nd, 3rd and 4th proximal ends show hyperostotic and slightly porotic lesions.   The distal phalanges are within normal limits.

The left femur, lower end, shows hyperostosis and porosity corresponding with similar processes on the tibia, most marked on the external condyle and tuberosity respectively. Hyperostosis and probably ankylosis of fibula to end surface of external tuberosity (broken).

The right knee is within normal limits.   Both feet are within normal limits, with the possible exception of one left metatarsal which has periarticular hyperostosis on the proximal end.

Right Hand.   X-ray shows the proximal ends of the metacarpals 2 to 5 to have slight rarefaction and irregularity of striae.   Shafts—wide obscure striae.   Distal ends are totally hypertrophic, thickened at the epiphyseal line which is in part absent, especially in metacarpals 1, 2 and 5. Irregular porosity in m.c. 3–4, definite striae lead to articular surfaces, especially in m.c. 3. First phalanges, proximal end, thickening of all articular edges, mixed in ph. 3 with porosities and circumferential thickening. Hyperostoses on ph. 4.

Chronic hypertrophic and necrotizing arthritis.

Gorilla (*Gorilla gorilla*). ♂ B-1,991. Td. Old animal. (Figs. 16, *a–k*)

Right glenoid—jugular joint is porotic, eburnated and flattened so that it was probably a loose joint. Right canines and first incisors distorted and alveolar processes partly absorbed. Left side of jaws must have been the effective and employed side.

The 2nd and 3rd cervical vertebrae show anterior and lateral lipping with marked porosity of the 2nd and 3rd interfaces and slightly of the 3rd and 4th interfaces. The 5th and 6th vertebrae show porous articulating faces but the shape is within normal limits.

The first and second thoracic bodies are narrow with marked dentoid lipping with some but no great porosity. These bones were in such close apposition and so worn that the intervertebral cartilages could not have existed. All the vertebrae have a smoothing off of their faces as if the cartilaginous tissue had become too thin and they fitted too close together. The lateral processes of the especially mentioned vertebrae were roughened and porotic.

The posterior ends of the first ribs are hyperostotic and porotic, indicating that an arthritis existed at this point. The second is very slightly affected. Below these the vertebrae are negative to the 12th thoracic, which shows superior lipping and possibly an increased porosity of the anterior third of the body. The 12th vertebral body is wedge-shaped and its anterior face with that of the 13th is worn down. The first and second lumbar vertebrae show marked dentoid lipping with porosity of the interarticular faces. These lippings are almost interlocked but there is no eburnation.

Right Hand. The carpus is within normal limits. The 2nd and 3rd metacarpals and first phalanges show marked porosity and subluxation. The 1st digit and 1st phalanx show porotic destruction of the distal end. The 4th digit, 1st phalanx articulating surface is entirely destroyed with a new false joint. The 5th digit, 2nd phalanx proximal end is entirely destroyed by an inflammatory type of change.

The left hand is not complete. The 4th metacarpal shows porotic destruction with corresponding change in the adjacent end of the 1st phalanx. The distal end is similar.

The right foot shows very slight change except in the 1st and 2nd digits. The distal half of the 1st phalanx is gone and there is a rough granular false joint with correspondingly rough proximal end of the terminal phalanx of the 1st digit. On the 2nd phalanx the proximal end is rough and the articulating surface is destroyed.

The left foot shows eburnation of astragalus and scaphoid with subluxation. The 5th digit, distal end of 1st and proximal end of 2nd phalanx are porotic. The 2nd digit, distal end of 1st, both ends of 2nd and proximal end of 3rd are porotic and hypertrophic.

X-rays. *Teeth.* There is destruction about the roots of lower m. 2 both right and left. Rarefaction of adjacent bone. All the visible parts of the alveolar processes of both jaws are somewhat rarefied.

Spine. First thoracic and adjacent ribs. Body shows porosity, irregularity of cortical line and hyperostoses on the surface. Articulating lateral processes slightly hyperostotic. The head of the rib is irregular by hyperostosis and rarefaction. Lateral picture of cervical and 1st three thoracics shows wedging of 6th and 7th and possibly 1st thoracic. Some thickening along the articulating edges, distinct irregularity of striae, also thickening of the posterior articulations where the articulating appositions were not specially disturbed.

These findings compared with the black and white picture seem to justify a diagnosis of hypertrophic and ulcerative spondylitis.

Hand. Metacarpals—proximal end, rarefaction only. Metacarpal shafts slightly thinned, median striae broken. Metacarpals—distal end, distinct rarefaction, especially seen in m.c. 3 and 4, porosities on the ends of m.c. 2 to 5 inclusive. Irregularity of striae around curve of articulating edge. Flattening of head of m.c. 4. Phalanges—proximal 2 rarefaction and lateral hyperostosis, proximal 3 marked porosity and rarefaction; no articulating membrane could be left. Ph. 4—porosity within the bone end, thickening of the delicate articulating edge that must have lain on the mushroomed end of m.c. 4. Ph. 5—osteoporosis and distortion.

The radiographs of this left hand suggest human ulcerative or atrophic (rheumatoid?) arthritis.

Gorilla (*Gorilla gorilla*). ♂ B-1,730. Td. Wild

This animal seems to correspond in skeleton to 45 human years. The left upper canine is missing and the right glenoid cavity is flattened out. Many bones, but especially the skull, show much demineralization—a very noteworthy finding in a wild animal. One can wonder if it is the result of metastatic calcification toward the healed fracture of the right leg and other places of hypercalcification.

Right Leg. Healed complete fractures of tibia and fibula just above middle. Hyperostotic false ankylosis at the malleolar ends of both bones, this being periarticular growth. There are some slight overgrowths around tibial tuberosities and head of fibula. Conditions in this leg may be entirely the result of injury.

Cervical Vertebrae. Hyperostotic and ulcerative changes from first to fourth. Thoracic—lipping and partial false ankylosis first and second. From the tenth thoracic to and including first sacral—lipping and wedge-shaped because the anterior height of the vertebral bodies is distinctly less than that at the cordal canal. Lumbar—second, third and fourth with the first sacral, complete false ankylosis. Posterior part of vertebral and costal articulations within normal limits. Both pedal complexes and especially calcanea, external tarsals and one pisiform, show hyperostotic periosteal roughening without ulceration.

Gorilla (*Gorilla gorilla*). ♂ B-1,430. Td. Plains variety—wild

Anatomist's Notes. All epiphyses united including those of the clavicle. All sutures of cranium and face united, no lipping of the long bones. Symphysis quiescent and rim complete. Probably equivalent to 30–35 human years.

Superior lipping, first thoracic, inferior and superior lipping, 8th, 9th, 10th and 11th thoracic. Superior lipping 3rd lumbar.

Gorilla (*Gorilla gorilla*). ♂ B-1,417. Td. Wild

Anatomist's Notes. Absence of all lipping indicates under 35 human years. Clavicles suggest over 28. Lapsed union in sacrum may be ignored. Symphysis not quiescent. Probable age about 32 human years.

Although anatomist has written that there is no lipping, observer finds this on the inferior edge on the 1st, both edges of 2nd and superior of 3rd thoracic and inferior edge of 4th lumbar, and believes the disease as great or greater than others of the same age stratum so that the bones should be considered pathological. Anterior surfaces of all vertebral bodies unusually rough and pitted but not necrotic.

Gorilla (*Gorilla gorilla*). ♂ B-1,408. Td. Wild

Anatomist's Notes. Character of scapula indicates a position slightly older than B-626 and than 1,416—17–18. Symphysis completely quiescent. Muscular development not great. Spondylitis of the vertebral column, condition of the skull, all comparable with human year 35.

Superior lipping first sacral left side; a very little on right, inferior lipping 4th lumbar. Calcification of the anterior parts of the intervertebral discs between the thoracic 1st and 2nd, 3rd and 4th, 4th and 5th, 10th and 11th, 11th and 12th. Minor double lipping between the pairs 7 to 12, inclusive. More on the right side and some about the costal insertions both sides. Fairly widespread spondylitis.

Gorilla (*Gorilla gorilla*). ♂ B-1,407. Td. Wild

Anatomist's Notes. Lipping of the scapula makes this animal older than any other of our specimens (this statement and similar ones doubtless refer to opinions on material studied up to time of receipt of this one. Ed.). Symphysis shows secondary erosion. Teeth greatly worn. Age probably about 55 human years.

Lipping on the inferior edge of the last thoracic vertebra, extensive lipping and hyperostosis of all lumbars. Complete false ankylosis of the last lumbar and first sacral. The posterior arch is not involved in any. Superior sacral apophyseal junction with last lumbar—hyperostosis. Intervertebral surface of last thoracic is porotic and of necrotic type on one side with corresponding necrosis on the upper face of the 1st lumbar. Sacrum possibly rotated slightly toward the left and away from the above necrosis between the vertebrae and in the direction that it would be fixed by the hyperostosis, which is great, of the lumbar bodies.

This skeleton is regarded as from the oldest animal available to the Haman Museum. The arthritic lesions are clear but not so great as in other skeletons estimated to come from younger animals, nor are they essentially different in pathological type. (See Gor. B-1,731.)

Gorilla (*Gorilla gorilla*). ♂ B-1,731. Td. Wild. (Figs. 17, *a–k*)

Animal's bones suggest that they are equivalent to those of about 35 human years. The findings suggest the existence during life of an infectious process.

Third and 4th cervicals, respectively inferior and superior edges, lipping with necrosis of the articulating faces—both changes are distinct. Third and 4th, spongy bodies with necrosis of articulating faces. Left foot-ulcerative necrotizing arthritis of proximal 1st metatarsal, of 2nd metatarsal and of 1st and 2nd phalanges of 4th toe. Right foot-ulcerative necrotizing arthritis of distal end of 2nd metatarsal and 1st and 2nd phalanges. Right hand, same lesions distal end of 3rd finger, 1st and proximal end of 2nd phalanx. Left hand, 5th finger, same lesions distal end 1st and proximal end of 2nd phalanx.

X-rays. Left Hand. Metacarpals proximal negative. Mc. distal-epiphyseal line diffuse, almost destroyed, some demineralization, striae irregular and distortion of the articular surfaces. Slight vacuolization in m.c. 2 and 3. Phalanges, 1, demineralization, irregularity and paucity of striae; shaft thin, bulging of the distal head of ph. 2 and 3, destruction of the distal and proximal head of ph. 5.

Right Hand. M.c. proximal, nearly normal. M.c. distal, porosity as on left hand. First phalanges comparable to phalanges on left hand. Distal 4th, porotic and ulcerative with corresponding porosis and ulceration on proximal head 2nd phalanx 4th digit.

Left Foot. Slight porosity metatarsal 1. proximal, where either a traumatic fracture or pathological separation of plantar protuberance has taken place. Mt. 2. to 5., inclusive, distal absorption, distortion, rarefaction, almost complete destruction of epiphyseal line.

There appear to be varying grades of hyperostotic arthritis and absorption at the ends of many bones, somewhat like osteomyelitis and therefore might be suggestively like rheumatoid arthritis. Changes in the left foot resemble early internal osteitis, possibly a forerunner of rheumatoid osteo-arthritis.

Right Foot. Second digit dorsal. Mt. thickening of proximal end and shaft, necrosis distal end. First phalanx, ulceration and swelling with sphacelus formation. These radiographs suggest very early inflammatory arthritis.

Practically all joints of both feet and hands other than mentioned show periosteal hypertrophy to a mild degree (only those noted were ulcerated and showed necrosis that could be sequential to actively inflammatory arthritis and almost certainly not incidental to multiple injuries). The necrotizing character of the spondylitis is consistent with the thought of rheumatoid arthritis.

ANALYSIS OF SPECIMENS ARRANGED ACCORDING TO ZOOLOGICAL ORDER

Analysis of the above material is begun from the lower end of the zoological tree, attention being directed to the salient features of the articular changes in terms of the anatomical and zoological peculiarities of each of the groups of animals.

MARSUPIALIA. This, the lowest of the mammalian orders to present instances of arthritis, is represented solely by the Macropodidae or kangaroo group. They are unusual in the development of their posterior skeleton, in marked forward inclination of many vertebral spines, marked angulation of the posterior limbs which are plantigrade, and the delicacy of the anterior extremity and head. Their progression is saltatory, under the power of the long foot and heavy tendons, assisted by the caudal balance. The jolt upon reaching the ground is adjusted for by the pliable feet and extensive angulation of the knees. There would be a pull upon the acetabular ligament and possibly a backward rocking of the ilio-caudal joint and of the lumbar and caudal spine.

They are almost exclusively herbivorous. Despite marked sensitivity of the respiratory tract and the frequency with which mycotic disease has occurred in them, bone disease is uncommon.

Two examples of this family, kangaroos, have distinct cervical lesions of hypertrophic and corrosive character leading to destruction of normal articulation. The wallaby had lower thoracic and sacral degenerative spondylitis. The kangaroo is the larger and jumps more than the wallaby. Its lesions are clearly cervical. As the animal jumps its head is held erect and stiff. Here is a suggestion that there may be a relationship of locomotion, leverage and rigidity to localization of arthritic changes. May it be, that in this animal, fixation of cervical joints is essential to balance, at the same time requiring great strain to be felt here? Restriction of circulation may also accompany contraction of cervical muscles and stiffening of ligaments.[3] Similar things will be seen in other orders.

In this study PROBOSCIDEA are represented by one elephant. His lesions require no comment but their widespread distribution deserves emphasis. Perhaps it is desirable only to note here that he had low-grade inflammatory arthritis associated with the very minor grade of captivity demineralization.

ARTIODACTYLA are represented by five families which will be discussed separately and then compared and contrasted. The single hippopotamus is included, although with limited lesions, because it was surely of truly wild origin, having been shot in the wild by the Theodore Roosevelt Expedition and presented to the National Museum.

In the SUIDAE arthritis is found in the spine and extremities. Two of the cases had porotic lesions in the vertebrae although at different localities.

A single giraffe had mid-thoracic hyperostosis on the anterior surface of the vertebrae.

CERVIDAE, the deer proper, supplied four cases. Spondylitis becomes more widespread, but better developed in the lower thoracic than in the upper or lower sections of the spine. It cannot be stated whether hyperostosis or porosis predominates. Extremity joints have a moderate number of lesions, notably in the elbow and knee complexes where porosis exceeds hyperostosis.

The BOVIDAE are represented by five large groups—the yaks (6), Indian buffaloes (2), goat antelope (1), antelopes proper (2) and the goat-like moufflon (1). There are too few of each to furnish sufficient material for contrast but sufficient similarity exists in the family to permit general description.

Spondylitis, while found throughout the spine, is commonest in the upper thoracic where it may be hyperostotic or porotic or both in the same case. The degree of arthritis is greater in the posterior than in the anterior limbs. The posterior part of the vertebral bodies is affected only in the thoracic section. Hyperplastic change was quite prominent in the apophyses and articulating processes of the Bovidae and in some cases, notably the Leche antelope, at the costal articulation principally. The lesions are more extensive in hypertrophy and more advanced in porosis, eburnation and luxation. Although many records were made for the front legs, the lesions were chiefly hyperostotic and usually of minor grade.

The number of lesions in the extremities was only sufficient for analysis in the shoulder-elbow and hip-knee complexes. Hoofed and horned animals give more instances of luxation associated with arthritis than does any other great group, a finding strongly suggesting its relationship to torso bulk and support. Anteriorly the shoulder joint showed hyperostosis, porosis, eburnation and subluxation about equally while the elbow complex was more often porotic. In the posterior limb, porosis greatly outfigured hypertrophy and there was more subluxation and eburnation.

---

[3] That this reasoning may not be generally applicable is fully realized, but nonetheless may be helpful. The kangaroo's tail, constantly subjected to strain and pressures in course of its functions, has not shown arthritis beyond what was obviously due to local injuries.

The ARTIODACTYLA are treated as an order in Chart I and the results are more striking than in the other groups. The cervical stretch is little involved (in one animal only, the yak) but much and severe arthritis is found from the 1st to the 7th thoracic segments after which involvement declines but remains at about the same degree throughout the rest of the spine.

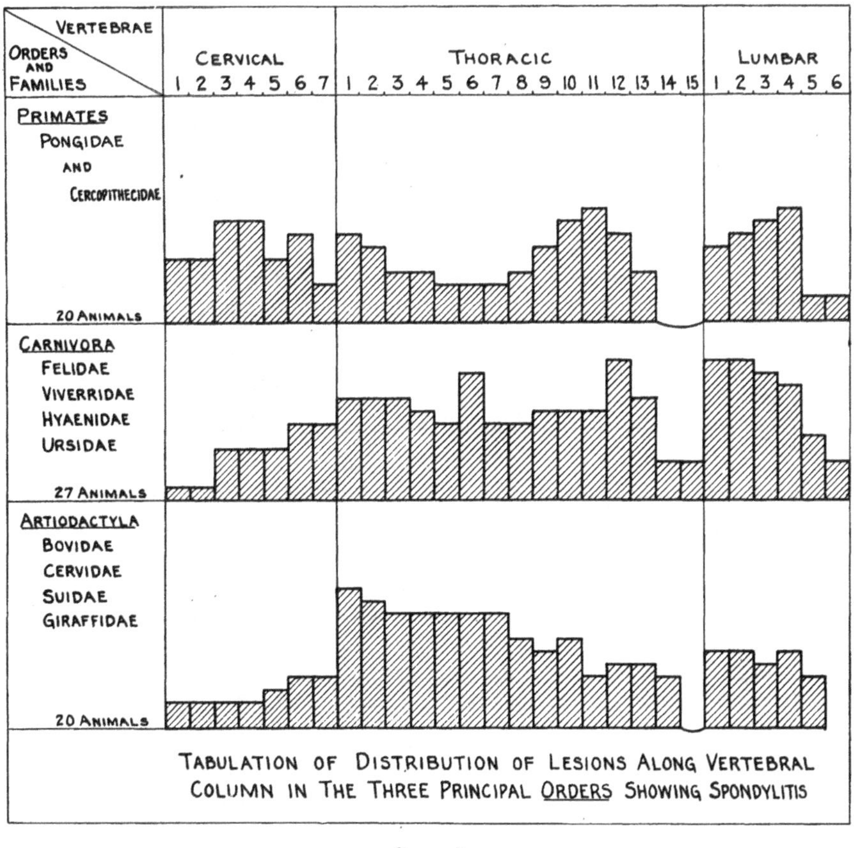

TABULATION OF DISTRIBUTION OF LESIONS ALONG VERTEBRAL COLUMN IN THE THREE PRINCIPAL ORDERS SHOWING SPONDYLITIS

CHART I

As can be seen in Chart II this order is made up of 12 BOVIDAE, 4 CERVIDAE and 3 SUIDAE. Examination of this chart reveals at once that the bovine family has influenced the ordinate charting of the incidence of spondylitic involvement and location. To evaluate this properly it is to be recalled that in order to find these cases, 87 BOVIDAE, 67 CERVIDAE and 18 SUIDAE were examined, so that respectively 13.8 per cent, 5.9 per cent and 17.2 per cent of the specimens studied showed spondylitis.

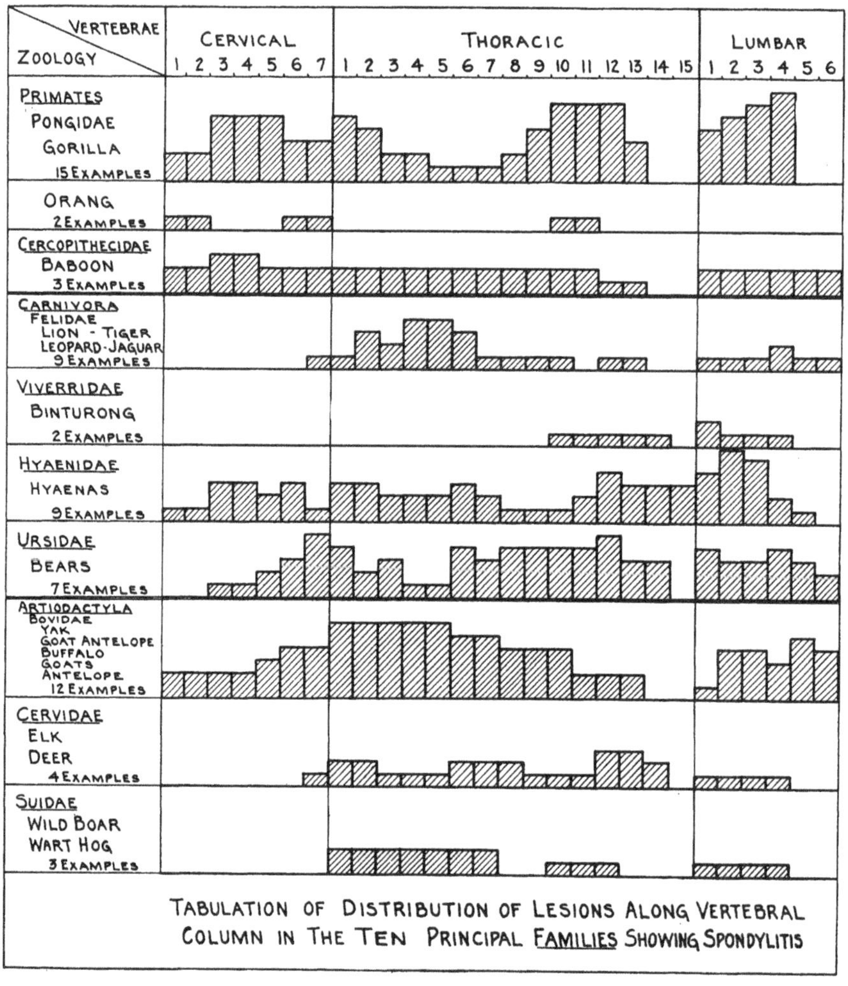

TABULATION OF DISTRIBUTION OF LESIONS ALONG VERTEBRAL
COLUMN IN THE TEN PRINCIPAL FAMILIES SHOWING SPONDYLITIS

CHART II

Analysis of the data as a whole reveals that thoracic spondylitis and arthritis of the femoropelvic and femorotibial joints are quite pronounced and appear to be characteristic for this order. Anatomically the lesions are of all sorts but those in the spine and those in the hind legs differ slightly in quality. Spondylitis is more hypertrophic, posterior limb arthritis is more porotic, polished and inclined to luxation. This general picture is to be contrasted with that found in the Carnivora.

Mention has been made of the cervical localization of the arthritis in the marsupial and the possible effect of stabilization of the neck in locomotion. Extending this thought to the ungulates, it would appear that, if the reasoning for the effect of tonic control of the spine upon limitation of nutrition and consequent pathology be acceptable, the greatest stiffening of the vertebral column would be at the lower end of the neck and the upper half of the thoracic column. This appears to be probable since the function of stabilization for the chest falls upon the antero-superior and lateral thoracic muscles binding the scapula to the side and moving the leg forward. Why the posterior extremity should suffer more than the anterior is not clear. Perhaps the jolt force of the step is absorbed by the posterior joints, damaging their nutrition and facilitating the localization of a causative agent more so than is the case with the fore limb. To these features must be added the weight of the huge belly barrel that must be supported by anterior and posterior spinal muscles, which weight would be directed backward as the animal walks or runs. While it may be accepted that the forward limb contributes the major stabilizing and upward thrust in locomotion, the major force in progression seems to come from the rear leg upon which is thrown the recoiling weight of the whole body.

PERISSODACTYLA are represented by a tapir and a wild horse. Their lesions are entirely vertebral and too few for comment.

The carnivorous mammals are introduced by a single case in a sea lion of the PINNI-PEDIA. Although its lesions are few and affect only the jaw and hip complexes, the case appears worthy of putting on record.

CARNIVORA proper are represented by the bears, the hyaenas, the viverrines and the cats. Canines, mustelines and raccoons, of which a considerable number has been available for study, show no lesions. Affected specimens represent the varied characters of the order in that some are widely distributed over the world, like the bears, while some are restricted in habitat, as the hyaenas; some are toe-walkers like the hyaenas, while others, as the bear, are almost complete plantigrade; some are strict meat-eaters like the cats, while the bears are the most herbivorous of the lot. However, they differ from the foregoing group, the ungulates, in the greater proportion of leg to trunk; none of them walks on stilts like the deer.

Eight specimens came from the URSIDAE and these were from animals with habitats all over the world. Four of them were found in the autopsy room of the Zoological Garden, yet even not suspected before death of having such severe articular disease.

Hyperostotic lesions are the usual findings in the bears, porosis and eburnation being inconspicuous. Hypertrophic changes occur in the lower cervical and upper thoracic regions, less so in the middle thoracic, while from the 6th thoracic to the sacrum the lesions are considerable and about uniform. The anterior extremity joints show a moderate grade of hyperostosis but the posterior limbs are slightly affected, though more so than in other carnivores.

Hyaenas, forming a family of their own, and limited to Africa, present a very striking picture and one that may be important. So serious is their arthritis that the moralist might look upon it as a reward for the animal's nasty temper, his slinking ways, his craven spirit and his habit of eating decayed flesh. These beasts accept and live well in captivity, even up to twenty-five years. Anatomically they are peculiar in possessing longer forelegs and shorter hind legs, in proportion, than any other ground carnivore. The fore-extremity resembles, in the obtuseness of the angles made by the joints, that of the horse, while the hind leg approaches that of the feline in arrangement. The vertebral bodies are like those of other carnivores, but the spines are quite like those of the herbivores. The animal progresses by the two legs of each side moving together, *i.e.*, a pace, but how he runs, jumps or stops is not known well enough for comparative comment. The musculature of the neck and anterior chest is enormous, seemingly well out of proportion to the moderate or even scanty muscles of the hind half of the body.

Only the skull and the phalanges are spared from arthritis in the hyaena. Nine out of sixteen skeletons had widespread lesions. Spondylitis is best seen on the 3rd to the 6th cervical, the 1st to the 6th thoracic and 11th thoracic to the 3rd lumbar, as hyperostosis, marginal necrosis, lipping and ankylosis, the last rarely complete, however. Where osteophytes are extensive, the intervertebral disc must often have disappeared because wedging of the bodies and exposure of the cancellated bone were prominent. Posterior lesions were not common and were of moderate grade. Arthritis of the anterior extremity joints was very pronounced in the hyaenas and lesions in the hind limb practically negligible. All kinds of lesions were found; hypertrophy probably predominated but ulceration had certainly existed on and around the articulating surfaces of some cases of shoulder and elbow-complex involvement.

VIVERRIDAE are represented by two binturongs (*Artictis binturong*, S.E. Asia). Their arthritis was limited to the lower thoracic and upper lumbar vertebrae as hyperostotic lippings without porosis. These are short-legged, heavy-bodied animals that jog along with a rapid walk or half trot. Their pathological reaction is less marked than that of other carnivores but occurs in similar places.

The cat family, FELIDAE, presented arthritis only in its heavier members, small ones like the ocelot and lynx not being found affected. Note can be made here that the great family, CANIDAE, assigned in zoological systems near to the two families just discussed, gave no cases of arthritis.

Cats are heavy bodied beasts in which the weight distribution appears more nearly equal on fore and after leg, a physical character possessed by all carnivores that offers a contrast with the other animals of this study. This family is nearly digitigrade in the forefoot and half plantigrade posteriorly. It contrasts with the bear in this for all bears use more of the full palmar surfaces of all four feet. This difference may assist somewhat in understanding the differences of arthritis in the extremities of the two.

The characters of arthritis in the FELIDAE are fairly distinctive. The spinal column is changed by hyperostosis along the anterior vertebral faces, to which may be added a low grade of interosseous porosis. Joints of the appended skeleton that have the greatest arthritic changes are those of the elbow complex where hyperostosis, ridging of articulating surfaces, porosis and eburnation may be found. Subluxation is not common; it is rare in carnivores. Arthritis in the posterior extremities is almost absent; one joint only was found affected.

Chart I shows the plotting of lesions in the spinal column of the Carnivora, as a whole order, according to the number of times the various joints were affected. It appears that arthritis becomes more prominent posterior to the cervical vertebrae, continues prominent along the thoracic spine to become quite marked in the upper lumbar region but almost to disappear toward the sacrum. High points are the 6th and 12th thoracic, 1st to 4th lumbar joints. These appear to be the regions of greatest tension of the spine. The upper chest bones and muscles form the stabilizing organs for the forward limbs. Why the low thoracic joints are diseased is not understood but the lumbar region in the Carnivora is a place of great activity since it is quite movable in nearly all actions and the abdominal and pelvic muscles insert deeply all around it.

If these Carnivora be separated as to family, as in Chart II, it is found that three important ones yielded almost the same number of cases of arthritis, Felidae 9, Hyaenidae 9, Ursidae 7.   Viverridae with two examples may be dismissed by referring to them as too few for comparison.   The general feline family has arthritis principally from the 2nd to the 6th thoracic and but slightly elsewhere.   The bears have the most prominent spondylitis, in the Carnivora, in a nearly uniform manner from the lower neck to the sacrum; upper middle thoracic vertebrae are less affected.   The hyaenas have spondylitis all along the spine with greatest prominence in the low thoracic and lumbar sections.   Simple observation of this animal in life suggests that the spine below the shoulders is kept rigid most of the time while the head and neck are in almost constant motion.

Analysis of the data to learn the involvement of extremities shows that the anterior limb has the greater number of lesions, at least in the main joints.   In the bears only does one find the posterior limb joints diseased and also in them the carpal and tarsal phalangeal complexes are affected.

Nothing could be plainer than the severity of the morbid process and number of joints affected in the anterior extremities, and the paucity of arthritic lesions in the posterior limbs.   Not only is the number high but the degree of pathological change is noteworthy. Bears appear to have a greater number of ulcerative lesions while hyaenas and felines have more sclerosis, porosis, eburnation and ridging, which happen to be more prominent in the shoulder and elbow complexes, than in any other order.   The peaks of disease at three foci in the spine and in the elbow and shoulder groups form a distinct panel for the Carnivora.

PRIMATES.   Now entering the highest mammalian order there is met a diversified group of quadrupeds and potential but undeveloped orthograde.   Their backbone retains the carnivorous form except in the anthropoid apes where certain human similarities appear.   One would expect that the cervical and upper thoracic spines would be developed in order to supply the head with powerful posterior muscles and a strong ligamentum nuchae; such is the case.   The greatest skull bulk, cervical musculature and thoracic posterior spines are found in the gorilla.   In this and other related Pongidae the foramen magnum is placed about the middle of the under surface of the skull, whereas in the Cercopithecidae it becomes, as one goes down the zoological scale, placed more and more near to the rear end of the skull.   Only in the baboon are the skull dimensions, ligamentum bulk and power of cervical muscles comparable to those in the Anthropoidea and incidentally to the carnivorous quadrupeds.   The lower spine is like that in the quadruped, perhaps most like the hyaena, the gibbon alone having a suggestion of the curves seen laterally in the human spine.   Shoulder, thoracic, arm, dorsal, abdominal and hip musculatures are comparable in all higher Primates.   The thigh and leg muscles are distinctly different.   Progression of Primates, gibbons excepted, is quadrupedal, yet all have the possibility of orthograde stance, thanks to the upward rotation of the pubis and direction of the midline of the pelvic acetabulum.   Larger anthropoids, especially the gorillas, walk upon the knuckles of the hand and the soles of the feet.   The joints of the fingers that are used are those between the first and second phalanges which are little affected by arthritis;

this is worst in the junction of first phalanx and metacarpal bones. If there are points at which skeletal fixation and osseo-muscular fixation must occur, it would be at the neck and the lower abdomen and pelvo-femoral region. Because of the occasional orthograde attempts, stabilization by abdominal muscles and lumbar spine would be necessary. If strain be effective in the localization of arthritis, these features of anatomy will be found very suggestive, and examination of the records finds them not entirely inconsistent.

The PRIMATES are represented in this study by gorillas, orangs and baboons; the single loris is of no consequence. Notable by absence are chimpanzees, gibbons, South American monkeys and hosts of Old World varieties. Those that are present in the list are, as are carnivores and herbivores, macrosomic, with a little disproportionate size of leg to body. Since the three genera shown in the table have so much in common it appears proper to discuss them together.

The cervical vertebrae present bones that are most seriously affected, showing ulceration, porosis, hyperostosis, subluxation, ridging and ankylosis. Below the cervical region, hyperostosis and false ankylosis are the usual findings.

Charts I and II represent the involvement of the vertebrae according to the number of times that the individual bones were diseased; the number of specimens yielding the material is indicated in the first column.

Examination of the Primates in Chart I reveals that the mid-cervical, upper thoracic, lower thoracic and midlumbar regions have the greatest number of arthritic joints, of which the lumbar part first and then the cervical seem to be most frequently diseased. Thoracic third to eighth have shown the fewest lesions. The four chiefly involved divisions of the spinal column are probably sections that participate most actively in the climbing power and stabilizing tension demanded by the mixed arboreal and ground life of the animals, to which may be added the strength in the neck needed to balance the huge head.

When this order is broken down into its families as in Chart II it is found that the gorilla has made the picture. The chimpanzee is missing; 24 of these animals were examined and no arthritis found; 89 gorillas were studied and 15 examples found; 41 orangs were examined and 2 arthritics found. Had as many chimpanzees as orangs been studied, perhaps arthritis would have been seen. However, it was very easy to find the disease in gorillas and it is naturally simple to explain their dominance in the charts. There is no striking lesson to be learned from the orang or baboon though it can be noted that the three cases of the latter, a definite quadrupedal African cyanocephaloid monkey, have a certain similarity to those in the Pongidae in the lesions of the cervical vertebrae.

This last point is also found in the data concerning the appended skeleton. Arthritis in the extremities of the gorilla again dominates but the baboon group also has it, while orangs do not.

Involvement of many manual and pedal bones and joints is characteristic for Primates or at least many more foci of arthritis are found in these animals, where the hand and foot differentiation is greater than in other zoological groups. Both gorillas and baboons have advanced porotic and hypertrophic lesions of metacarpal and metatarsal joints with their corresponding phalanges and in these latter bones also. Perhaps the porotic lesions predominate in the phalangeal joints of gorillas and baboons to a greater extent than in the joints of any other variety.

Gorillas 1,731 and 1,954 and mandrill 12,147 are, in the writer's judgment, so like the standards set for rheumatoid arthritis that this diagnosis is given for them, fully realizing that very few pathologists have seen such complete skeletons of man having this disease. It may be added that two competent pathologists have made the observation that these lesions are acceptable with what they know of human joints and by x-ray in cases diagnosed as rheumatoid arthritis. Radiographs of these are reproduced and discussed in the appropriate section.

Emphasis should be laid upon this primate material. There are twenty cases reported. Sixteen of them were in animals taken in the wild. Therefore systemic arthritis comparable to that seen in man occurs among Primates, to which order man belongs, in truly wild nature.

There can be added to the records of this article a communication from Dr. A. E. Hamerton of the London Zoological Society concerning their experience with arthritis over fifteen years. It is interesting that the observations there support the findings of the present study in that the groups with no arthritis are similar to ours save for one case in a macacus monkey. Dr. Hamerton writes:

"I can find no case of arthritis in any Rodent, Sloth, Anteater, Armadillo, Bat or Canidae. Japanese Monkey (*Macaca fuscata*), ♀, an aged animal, died from chronic bronchitis and emphysema. Suffered from deforming arthritis of all vertebral joints, causing kyphosis and rigidity of spine. Intravertebral discs and ligaments ossified, no serious damage to other joints. Hamadryas Baboon (*Papio hamadryas*) two cases, both males, aged 25 years, died from senile degeneration in cardio-vascular system and glandular organs. Suffered from deforming arthritis of all vertebral joints similar to the case above quoted. The large joints of the limbs were not affected. Indian Elephant (*Elephus maximus*), ♀, aged 38 years, died from gastric ulcer, showed marked erosion of cartilages of hips and shoulder joints, eburnation of underlying bones and inflammatory thickening of ligaments. European Bison (*Bison bonasus*), ♂, aged 16½ years, died from senile renal arterio-sclerosis. Senile osteoarthritis in both shoulders and hip joints and ossification of the attachments of surrounding ligaments and tendons. All bones showed extensive senile atrophic osteoporeosis. Cape Buffalo (*Syncerus caffer*), aged 20 years, ♂, died from haemorrhagic pancreatitis, showed kyphosis and lateral curvatures of spine. Vertebral discs and ligaments calcified—and bodies of vertebrae distorted by osteophytes, also osteo-arthritis affecting the knee, shoulder and hock joints of both sides. Tiger (*Felis tigris*), ♂, aged 15 years, died from multiple gastric ulcers. Both hip joints of this animal were disorganized by chronic osteitis with formation of numerous osteophytes in and around the joints."

GENERAL COMMENT ON ARTHRITIS AS A PHYSICAL CHANGE

In a survey of the cases cited and their zoological positions, it may be stated that arthritis comparable to the human varieties occurs in several widely separated varieties of wild animals. The most numerous lesions are found in the vertebrae, where lipping, fibrosis and calcification along the course of the anterior ligament and false ankylosis, represent the hyperplastic processes, and ulceration and porosis are of atrophic or inflammatory nature. Morbid changes are notably found on the edges of the bodies anteriorly but may affect the posterior half as well. Occasionally, as in the Primates, porotic and destructive change with collapse of the vertebral body may be seen. Lesions of the extremities, except in the Primates, are predominantly those of hypertrophic arthritic nature; some distinct destruction of bone ends was seen in felines and ruminants. Gorillas have presented two cases showing swelling and demineralization of small bones composing joints, with and without periarticular ostosis, that compare favorably on gross inspection with what is accepted as rheumatoid arthritis.

Cases Bear, P.Z.G. 11,800 and Baboon, P.Z.G. 12,147 offer characters that may assist in understanding the two great groups of human chronic arthritis. The lower thoracic vertebrae of the Bear are shown in Figs. 7A, B as illustrating hypertrophic degenerative osteoarthritic lesions with ankylosis. Externally, Fig. 7B, this is very clear. Longitudinal section of the bones, Fig. 7B, reveals how the falsely ankylosed bodies are tightly bound and the intervertebral disc has disappeared—all this with osteoarthritis. Figs. 13D, E are from the Baboon where cervical ulcerative lesions, best seen on the free face of the separated vertebrae, appear to be dominant yet have resulted in true ankylosis as indicated by the two bodies without any vestige of intervertebral disc. In these two cases the gross characters of what are often taken to be two essentially different diseases—rheumatoid arthritis and osteoarthritis—are comparable; one might think that differences of rate and intensity of pathology were responsible.

COMMENT ON ZOOLOGICAL CHARACTER

Data here presented indicate that many orders, from the lowest to the Primates, have arthritis in natural and captive life. Emphasis can be put upon this by reference to the many examples of gorillas that Dr. Todd obtained directly from Africa where the natives are rewarded for bringing bones to missionaries and animal collectors. Another example of articular change in a truly wild animal is that of the goat antelope, an exceedingly rare animal.

The varieties that have arthritis are macrosomic, at least more so in relation to their supports, than those that do not, as will appear shortly. Even if the size of the torso be great in all arthritis-bearing varieties, the extremity character is not always the same; the deer can be well contrasted with the gorilla. In arthritics, however, the legs may be thought to have an important part in the individuality of poise and gait, possibly more so than in those failing to reveal arthritic lesions. This of course directs attention at once to the jolt or strain incident to locomotion or feeding actions. All the potential functions of mammalian extremities may be found in the arthritis group, as also in the non-arthritic ones. This includes support, flexion, extension, rotation and digital grasping. It does not appear that these functions of themselves stand in any relationship to arthritis of the legs; perhaps the animal in which all functions are most developed, the gorilla, has more extensive and diversified lesions.

Probably the most important function of extremities is locomotion which does seem to have a relationship to arthritis. The greatest contrast in the panels of localization of arthritic change exists between Carnivora and Artiodactyla which for argument's sake can be repeated; Carnivora have arthritis prominently in the forelimbs and spondylitis all along the spine and more in the caudal half; Artiodactyla have conspicuous arthritis in the posterior extremities and spondylitis most marked in the cephalad thoracic spine.

Since it is possible to associate arthritis and locomotion, perhaps there is a relationship to jolt or pressure action; perhaps it is not unfair to recall that draught and circus horses have the disease and that flattening of rounded bone ends and lipping on their margins occur in late human life. If this thought be acceptable it would seem that the locations peculiar to these two contrasted orders, carnivores and the split-toed, may represent the muscle-joint combinations that bear the greatest strain of progressive locomotion. Possibly the fixation of the thoracic spine about the withers to act as a stabilization center for the shoulder complex, aided by the rigid fixation of the ligamentum nuchae on the upper thoracic posterior spines, may reduce the freedom of motion in the upper thoracic vertebrae of the artiodactyles, pulling on the intervertebral discs and perhaps also reducing the blood supply. The upper thoracic region seems to be the least mobile part of the spine of this order. Jolt action on the posterior limbs is not easy to explain, especially since the center of gravity of this animal appears to be just behind the scapula. Perhaps it is not so much the jolt from the earth as the tension exerted by the hind quarters in running that puts a strain upon the rear legs. Whereas the stop-jolt of locomotion and the upward thrust are felt most by the anterior legs and shoulders and some forward power is supplied by them, most propulsion seems to be derived from the rear legs that move forward, almost to the center of gravity, and must counteract the recoiling weight of the whole body.

Explanation for the Carnivora on the same basis would require the assumption that the lower thoracic and lumbar regions are put on greater tension and jolt and that locomotion impact is felt chiefly by the forelimbs which also provide propulsive force. As in ungulates the stop-jolt of the cat is felt by the anterior leg which, however, in the early phase of the step, produces a retarding effect; this must be true also of the ungulate. Most of the power of the hind limbs is used for propulsion, again against the weight of the anterior part of the body.[4] The construction of the erector spinae group and ligamentum nuchae is somewhat different in being less bulky in relation to the vertebrae and the action of the carnivore's back is certainly more supple than that of the artiodactyle. The functions of the anterior extremities of the carnivore are more numerous, especially in pronation and digital power, and it seems that greater use is made of the adductor group of muscles, including those from the cervical region. If jolt effect be related to extremity arthritis, one would have to assume that the forelimb of the Carnivora has more strain put upon it than would be the case for the Artiodactyla. In addition the musculature of the neck may play a part in the disease of cervical vertebrae.

CONTRAST OF ANIMALS WITHOUT ARTHRITIS

Many orders are missing from the list of arthritis-bearers. The most prominent are Lipotyphla, Chiroptera, Rodentia, Tubulidentata and Xenarthra. Prominent missing families are Cebidae, Hapalidae, Canidae, Mustelidae, Procyonidae, Tapiridae, Tayassuidae and Didelphiidae. It is most emphatically not intended to assert that only certain varieties have arthritis, while others have not. No figures of percentage incidence are given. However, when nine of fifteen hyaenas have arthritis, the hint is strong. When many hundred Canidae, rodents, bats, hedgehogs, sloths and armadillos are studied and no arthritis found, it may mean something. At once it will appear that arthritis-bearing animals are coarser, heavier animals than those that do not have arthritis; indeed, the rhinoceros and camel are the only very bulky beasts, of those studied, that do not appear in the arthritis list. This being noted early in the mustering of the data, small monkeys, bats, sloths and armadillos were especially well studied without discovery of lesions. Non-arthritics come from the same geographical areas as those having joint disease.

[4] John Manter, unpublished work.

ECOLOGY AND HABITS

No particular features of these subjects separate arthritic from non-arthritic varieties. Both represent the whole zoo-geographical world. Old World varieties predominate in the list but they do also in the available skeletal material. Herbivorous animals are represented by 56 (classing bears with this group because of their habit, not their zoological position); carnivorous 21; this corresponds fairly well with the availability of bones for study. While this distribution is clear enough, it should not be accepted finally or too literally. Degree or severity and locations are sufficiently different within the dietary groups to offer material for speculation and study, discussed, in suggestion at least, at other places.

The spread of arthritic lesions through the mammalian list makes it improbable that much would be learned from a study of zoology and ecology of individual groups. Perhaps localization and nature of the lesions themselves suggest that terrain and musculo-osseous motions may have some influence. Primates live on the ground and climb trees in a warm climate, probably well supplied with food and regularly appearing wetness. They run and jump much as the carnivores do and their varying habitats are not so very different. The ungulates offer many more habitats, demands for the search of food and self-adjustments to the terrain. However, there seems nothing in the " home life " of arthritics that demands among themselves peculiar changes of adjustment.

ASSOCIATED MORBID STATES

Can pathological tendencies offer any light on the matter?—unfortunately little, as the following indicates. Perhaps the most discussed morbid relationships of arthritis in man are rheumatism and focal infection. If one determine rheumatism as a recoverable arthritis without local sequelae, but with subsequent endocarditis, chorea, arterial obstructions and Aschoff bodies, one can state that no satisfactory evidence exists that this disease occurs in wild animal autopsies. The writer cannot answer for the gorilla in which so much arthritis has occurred. A hyaena and a bear, dead at this garden, and revealing systemic arthritis, were, however, sufficiently studied to permit the report that nothing acceptable as endocardial and endovascular hyperplasias or myocardial Aschoff bodies was discoverable. While domestic animals may have rheumatism, and this is extremely doubtful, its existence in wild ones remains to be shown.

Focal infection from the teeth can be assumed in many varieties, if one judge by disease in the teeth and jaws; this is not so evident in wild as in captive specimens. The denture of wild-shot animals is usually excellent, at least in so far as extensive caries and osteitis are concerned. Many menagerie bodies present apical dental infection but the principal changes are retraction of the gum and rarefaction of the alveolar section. Notable exceptions are, however, admitted, as in Papio 12,105 and Mandrillus 12,147.

The measurement of the importance of focal infection is by no means an established achievement of modern medicine. Most often the causative relationship of apical abscesses, cholecystitis, arthritis, subactive nephritis, prostatitis, et al., is assumed or presumed, although it must be admitted that the cure of one is sometimes followed by the cure or improvement of another. Focal infection, as seen by the writer at the Zoological

Garden, is exceedingly difficult of establishment. Local dental and alveolar necrosis has been followed by pneumonia, as has an abscess in a far removed spot. Dental infection and enlarged tonsils were associated with biliary lithiasis. For the most part there is a flagrancy of the original spot that might give rise to vascular embolism, characters that remove the morbid process from the category of focal infection which is connoted with a slower or cryptic relationship of bacterial localization and secondary effect. As a contribution to this study, the relationship of the teeth and arthritis was investigated by having made roentgen ray pictures of a number of skulls. Going as far as our finances permitted, 32 skulls were photographed, 16 from known arthritics, 16 from known non-arthritics— 8 of each were within normal limits, 8 of each showed apical lesions or absorptions or more extensive osseous disease. From these limited data, the jaw cannot be finally convicted as a source of the arthritic agent.

Search in these arthritics for pathological proclivities or panels that might bring them together, divide them into groups, or distinguish them from non-arthritics, has been fruitless. There are indeed morbid tendencies in animal orders—vulnerability of the Old World Primates to tuberculosis while the carnivore resists it; the rodent has many spontaneous tumors, the ungulate very few—but no one stands in comparative relationship with arthritis. There has been an attempt to associate tuberculosis with arthritis in human pathology but the thought has not received wide credit. It is interesting, and that is all, that apes and ungulates are susceptible to spontaneous tuberculosis and arthritis, but Old World monkeys, easily infected by the tubercle bacillus, have failed to show arthritis. Arthritics and non-arthritics do not occupy respectively any other distinctive differential morbid panels.

Since the endocrine system has been thought of in connection with arthritis and since the wild animal varieties having the disease might be considered as macrosomic by reason of pituitary, thyroid or thymic dominance, it is interesting to record their pathological panels in the glands of internal secretion. Abnormalities of these organs are extremely rare in Primates. Artiodactyla have suffered with hyperplasia of the pituitary with fatal results by erosion of the skull. The thyroid and thymus are very rarely pathologically altered in them. Carnivora, as especially the non-arthritic Canidae, have many thyroid abnormalities but few endocrine disorders within the cranium or thorax. The adrenals are never greatly abnormal.

CONSIDERATION OF DIET AND AGE

While dietary grouping of the affected beasts shows a preponderance of herbivores, the full significance of this would better remain for more information. More Primates and Ungulata skeletons were available for study. What is possibly of value is the extent and degree of morbid anatomy in the two great groups—those living chiefly on meat, those preferring vegetable food. In making this analysis, one is a bit confused by the serious affections of the hyaena in which beast the greatest arthritic development occurs. However, even including this single variety, lesions in carnivores are more hyperostotic and deformative than in vegetable eaters, where more porosis and ulceration appear even though combined with osseous overgrowth in the immediate vicinity. There can be little doubt of the differences in the distribution of lesions between Carnivora and Ungulata (*q.v.* pages 112–114) but there are certain similarities and differences between the two great herbivorous orders, Artiodactyla and Primates. Cervical disease stands out in the simians but not in hoofed and horned animals. Thoracic lesions appear in each but much more in the ungulate group. The lumbar and sacral regions are similarly affected. There is no panel for the extremities that accords in the two. Diet preferments appear to have no clear-cut relationship with arthritis in these groups.

Diets in the animals in the non-arthritic list, repeated for ease of reading—Canidae, Rodentia, bats, small monkeys, sloths and armadillos—are essentially the same as for those with chronic inflammation of the joints. Their morbid proclivities, perhaps too little known for the most part of them, seem to have no panels of similarity or dissimilarity that permit comparison or contrast.

Age is another factor that pertains to human arthritis in a supposedly causative manner. Unfortunately no dependable data are to be had upon enough specimens of wild animals to permit comparison with man. One can state that an animal is young, mature, or senile, that is about all. From menagerie records, the exhibition time of arthritics is quite suggestive. These range from 91 to 325 months, which is greatly in excess of general exhibition averages for the several varieties, and would permit comparison with maturity and high maturity ages of man when osteoarthritis is more common. This is true for varieties as antelopes and other jumping and running ungulates and for the slowly moving, more nearly plantigrade felines and bears. Gorillas have ages in terms of human years estimated by Todd upon the condition of the pubic symphysis, cranial sutures and epiphyseal ossification lines. Most of them belong to the fourth human decade when rheumatoid arthritis occasionally arises and when osteoarthritis often begins.

Arthritis was not found in any of the menagerie animals that could be assigned to an age that corresponds to adolescence or early maturity when rheumatoid arthritis of man is customarily found. As already indicated, all the skeletons were from animals of full growth.

A word might be given to life histories of non-arthritics—Canidae, Mustelidae, Chiroptera, Rodentia, Xenarthra, Lipotyphla and others of which fewer skeletons were studied. Perhaps these varieties fall into such a group because an insufficient number of skeletons was examined but one cannot go on rattling bones forever and it has been easy to discover arthritis in several other varieties, gorillas and hyaenas for examples. Moreover a reasonable number of skeletons of these non-arthritic varieties was examined and the results are quoted simply from the material.

So far as acceptable figures of life-years are concerned, data at the Philadelphia Garden would indicate that many examples of non-arthritics live well beyond the family average and thus fall into the age groups of arthritis-bearers. Anatomically the two groups are similar in some instances, quite different in others. Hyaenas, while not dogs, suggest them and have similar organs, functions and behavior to non-arthritic canines. Mustelines and viverrines are not greatly unlike. Chiroptera and Xenarthra are wholly dissimilar to any arthritic. Why the rodents are not bearers of arthritis is a puzzle; some have ungulate anatomy and habits and some carnivorous composition. Rodents, as an order, have a short total time-life when compared with many other groups. If a life-span represent the same biological process regardless of time, the rodents have opportunity for youthful infections or later sequelae and for senile irreversibility in tissue.   No arthritis has been seen in 173 animals.

BODY BULK

If there be one character that separates arthritics from non-arthritics it is body size. With the minor exception of the marsupials, arthritics are macrosomic whereas non-arthritics are on the whole small beasts.   Macrosomism may exist in the form of great torso and thin supports as in the Bovidae, or as stouter supports to a stockily built torso, as in the Felidae.   Body bulk and the greater incidence of luxation in the Artiodactyla is an association already mentioned.   Mention has been made of the jolt effect of locomotion and the significance of the abdominal barrel.   Canines and rodents present examples of animals that possess bodies and habits subject to both of these influences while the mustelines and xenarthra do not.   With exception of hyaenas, bulk weight, joint tension and gait jolt all stand in some relationship to the incidence of arthritis, it would seem.   Simple excess motion of one pair of extremities, as in the bats, has no comparison or contrast in others.

SUMMARY

The analysis of more than seventeen hundred skeletons and autopsies of wild animals reveals in the joints, changes that correspond with chronic arthritis in man.   Not only has this been discovered in specimens exhibited in menageries but also in material that was certainly in its proper wild habitat when killed.   The lesions of these " truly wild " animals are entirely comparable to those from captive specimens.   It is evident therefore that chronic arthritis occurs in wild nature.   (See goat antelope, Tibetan bear and gorillas.)

It has not been difficult to discover these cases and it has been reasonably simple to learn which varieties have the most conspicuous lesions; Anthropoid apes and baboons, Felidae, Hyaenidae, Ursidae, Bovidae, Cervidae and a few others.   Reversely it has developed that a number of groups, notably certain families of Carnivora (e.g. Canidae) and some orders like the Rodentia and Chiroptera, do not appear to have arthritis in the material of the study although a very considerable number has been examined.   The ease of discovery of the disease in hyaenas and gorillas should be emphasized.

No attempt is made to give percentage; it is only stated that 77 cases were accepted as arthritic in the study of 1,749 skeletons listed in Table I.

Diagnosis and description are based upon dried skeletons from museums or wet-cleaned bones at autopsy; in many cases radiographs have added some support.

Arthritis is best developed in the spinal column, there being some grade of spondylitis in nearly every case; it can be stated as the dominant articular localization.   Productive articular and periarticular disease is more prominent in carnivorous animals as against porotic and ulcerative change in herbivorous animals.   (Contrast hyaenas and gorillas.)

Arthritis of the extremity joints is found in all varieties but has a distinct quality in the Carnivora vs. the Artiodactyla.   While the morbid lesions agree with the conditions made in the foregoing paragraph, the localization is different.   Carnivora have more serious arthritis in the forelimbs and almost none in the hind limbs.   Artiodactyla have more conspicuous lesions in the hind limbs, although the forelimb joints are also often diseased.

Certain distributions of spondylitis distinguish the different great animal groups in such a manner that a relationship of function and lesion is suggested. It appears that where the greatest stabilization of the spinal column is demanded for the animal's locomotion, there the greatest degree of spondylitis arises. Perhaps jolt-shock and locomotive power have something to do with the localization of lesions in the forelimbs of carnivorous animals and hind limbs of the herbivorous ones. Lesions of the hand and foot complexes are more numerous in those animals, the Primates, in which greater specialization of evolution and function exists. In these animals both hypertrophic and rarefying arthritis is to be found. Next to them perhaps come the bears where a smaller degree of morbid change was found but in the zoological groups where the carpals and tarsals are less well differentiated, Ungulata, arthritic lesions are not conspicuous at these joints.

It is notable that the most conspicuous arthritis-bearers—gorillas, felines, bears, hyaenas, bovines, cervines and wild swine—are macrosomic. Many have very slender supports or legs that appear too small for the torso; bears and gorillas may be somewhat of an exception. Some varieties without arthritis have also this construction but many in which body-bulk and sturdiness of leg appear less in contrast—South American and Old World tailed monkeys, rodents, bats—are missing from the list. Macrosomism, absolute and relative, is conspicuous in the arthritics.

There is no relationship between the existence of arthritis and the place in the evolutionary scale assigned by zoological taxonomists to an animal. Perhaps the localization in terms of anatomic specialization is another matter. The gorilla has much cervical spondylitis and this animal, with the bear, has many lesions in the hands and feet.

Age in a wild animal is almost always impossible to state. With rare exceptions they could be classed as fully adult. Analysis of autopsy data on menagerie material of known exhibition age permits the record that arthritis-bearers had lived well beyond the average of their order and family. Non-arthritics also were found to be of average or over-average life.

There could be found no relationship of arthritis and zoo-geography, ecology, individual habits, pathological panels, diet, focal infection such as dental and alveolar disease.

Finally it is cautiously suggested that there is a strong similarity between human hypertrophic or osteo-arthritis and that in lower animals, as found in Bear, P.Z.G. 11,800, and between human rheumatoid arthritis and the cases of Gorilla Td. 1,731 and 1,991; others also exist and these are only examples.

PLATE I

Fig. 1. Goat antelope. N.M. 258,652. Right femoral head, anterior and posterior, illustrating chronic arthritis and periarthritis; comparable to human morbus coxae senilis. Page 87.

On left, Control of a normal bone from a similar antelope. (No available goat antelope for comparison.)

Fig. 2. Leche antelope. P.Z.G. 9,592. Page 88.

A. Lower cervical and thoracic vertebral column showing extensive hyperostotic growth in the anterior thoracic region and the costovertebral joints. On left, Normal vertebral column as control, from an antelope of the same height. (Photographs not on exactly the same scale but normal construction is clear.)

B. X-ray pictures of the above two spines. Thickening of the vertebral joints, delicate overgrowth on the anterior surface, which appears to be much more delicate than the black and white photograph would warrant one to expect. Destruction of the fourth thoracic vertebral body.

PLATE I

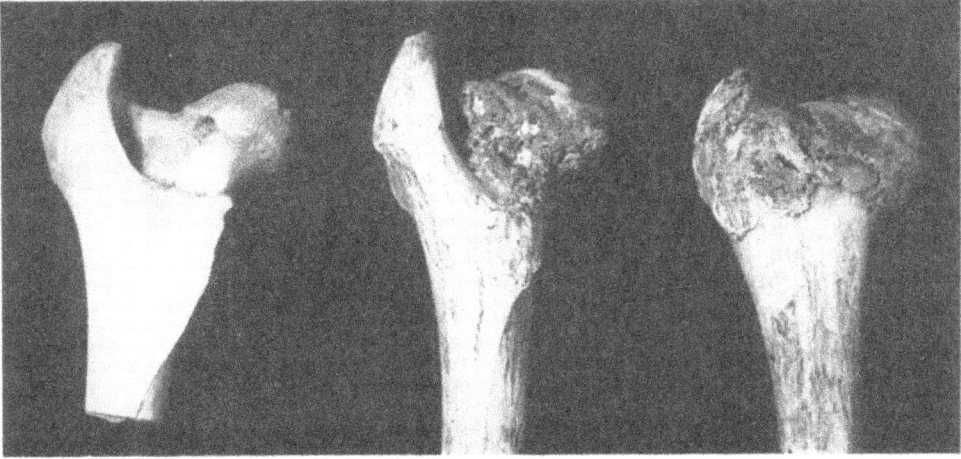

Fig. 1.

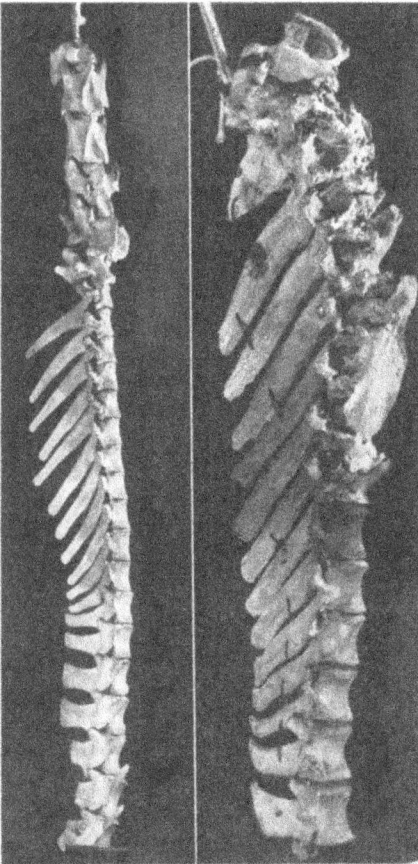

Fig. 2A.

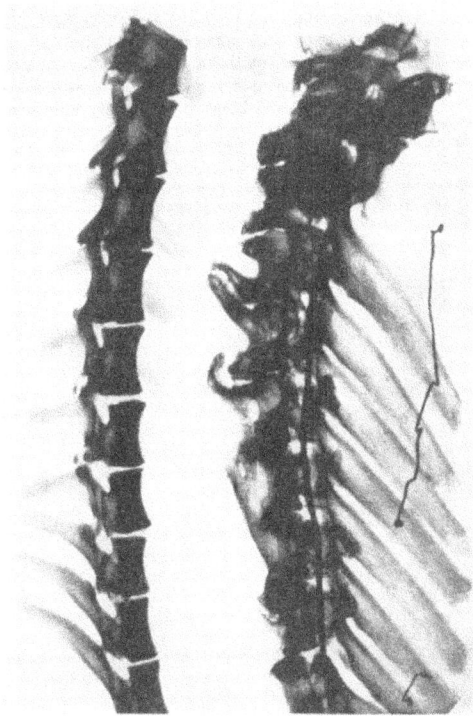

Fig. 2B.

PLATE II

Fɪɢ. 3. Yak. A.N.S. 3,078. Page 89.
 Femur and tibia on right and open knee joint on left, showing hyperostosis of epiphyseal regions and periarticular zone and early depression and erosion of articulating surfaces.

Fɪɢ. 4ᴀ. X-ray of the thoracic and lumbar vertebrae of a lioness believed to be entirely normal. Illustration to be used for comparison with other vertebrae with large bones. The internal structure of the vertebral bones is essentially the same in the Primates. Dorsal and lateral views.

PLATE II

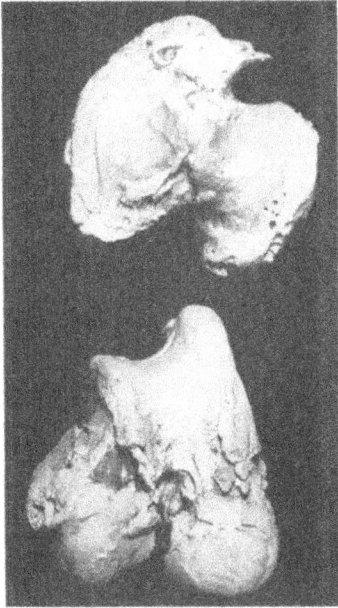

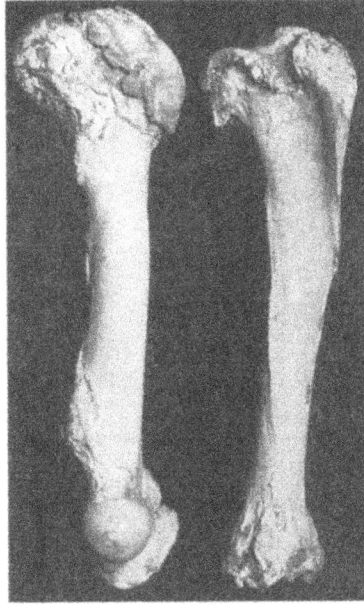

A

B

Fig. 3.

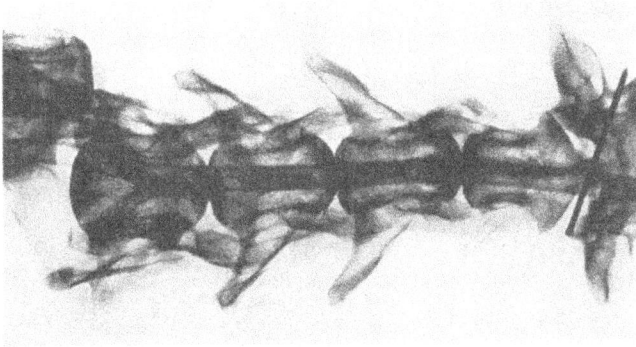

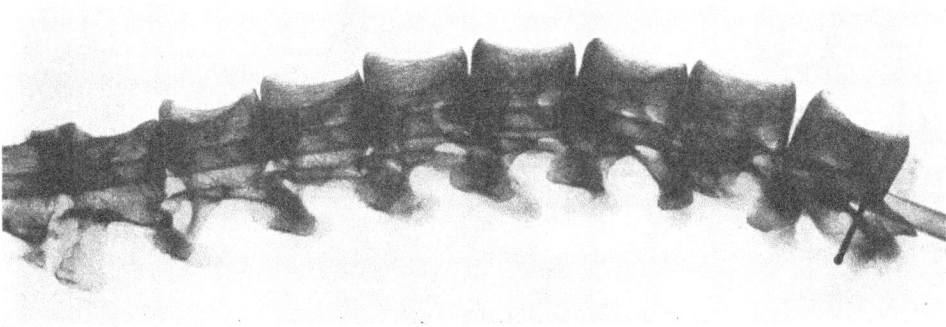

Fig. 4a.

PLATE III

FIG. 4. Spinal columns of Hyaena, A.N.S. 4,260 and A.N.S. 4,261, Alaskan Bear P.Z.G. 11,789 and Yak A.N.S. 3,078.

These illustrate different locations of hyperostoses and show the character of spondylitis deformans in the three kinds of animals, thoracic in yak, all along spine in others.

PLATE III

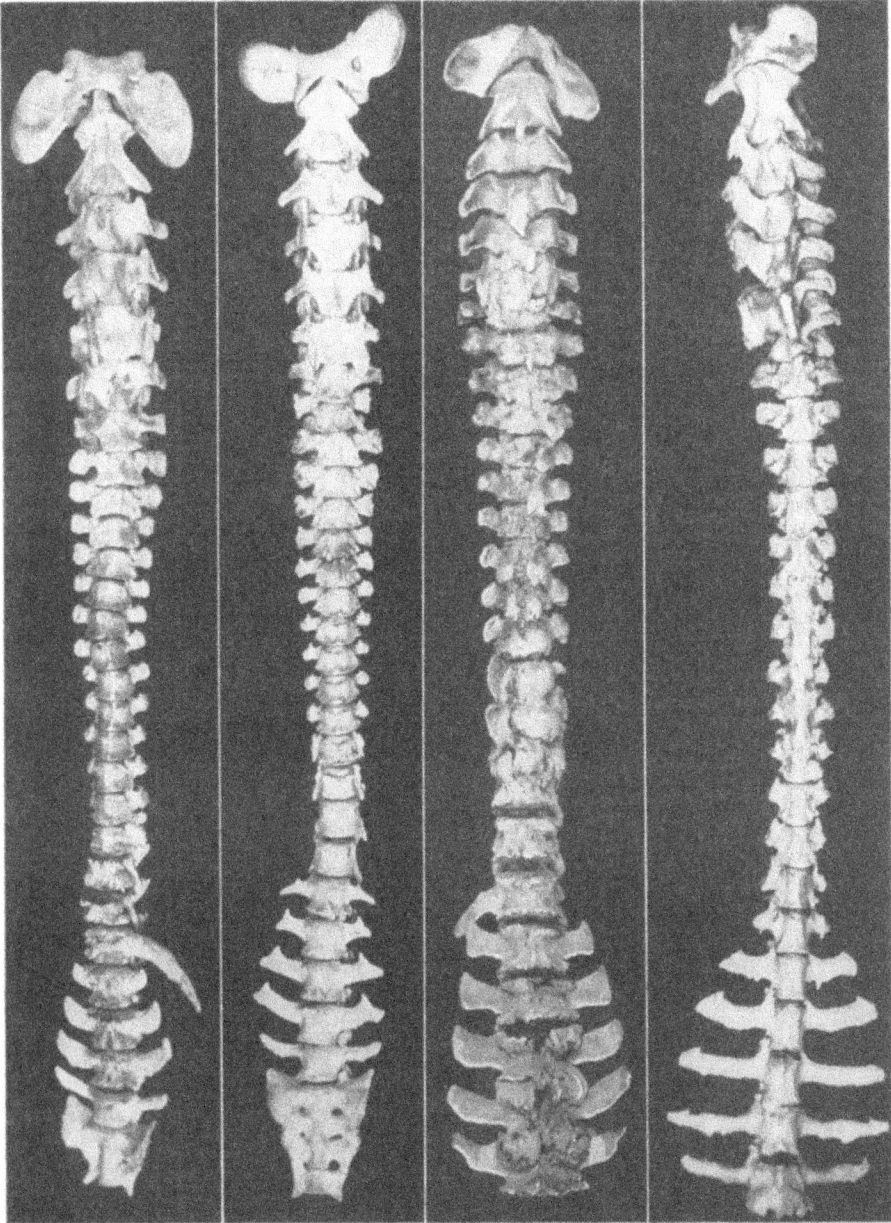

A.N.S.P. 4260.
Hyena.

A.N.S.P. 4261.
Hyena.

11780.
Alaskan Bear.

A.N.S.P. 3078.
Yak.

PLATE IV

FIG. 5.   European brown bear.   P.Z.G. 11,954.   Page 92.
   X-ray of lumbar vertebrae.   Irregular lipping and curving on all interosseous junctions; early
   wedging; density of cortical bone and posterior articulations; rarefaction of centers of bodies.
FIG. 6.   Black bear.   P.Z.G. 11,792.
   X-ray of jaws showing absorption about the roots of the canines and the incisors.   This animal had
   no arthritis.
FIG. 7.   Black bear.   P.Z.G. 11,800.   Page 93.
   A. X-ray.   The sixth to eleventh thoracic vertebrae with hypertrophic and ankylosing spondylitis.
   B. Show enlargement of the seventh and eighth vertebrae, viewed externally and in longitudinal
   section, showing false ankylosis that has destroyed intervertebral disc.

PLATE IV

Fig. 5.

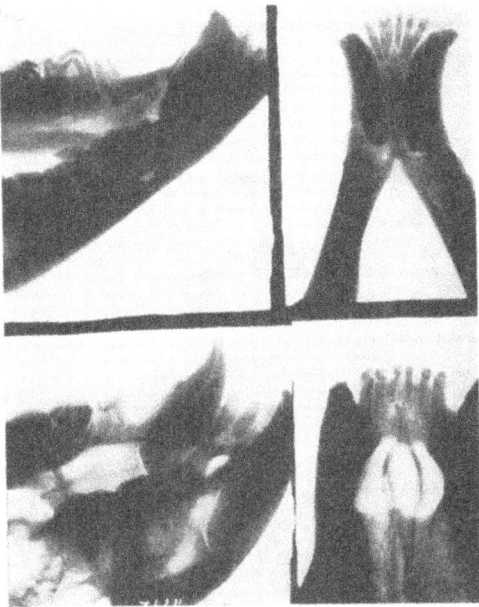

Fig. 6.

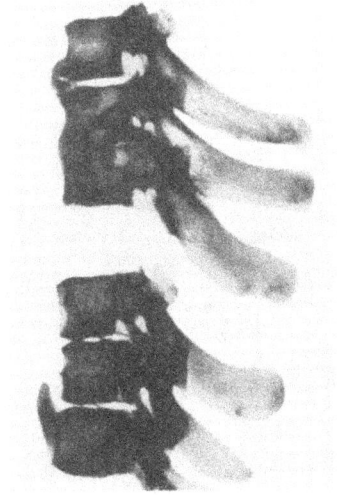

Fig. 7a.

Fig. 7b.

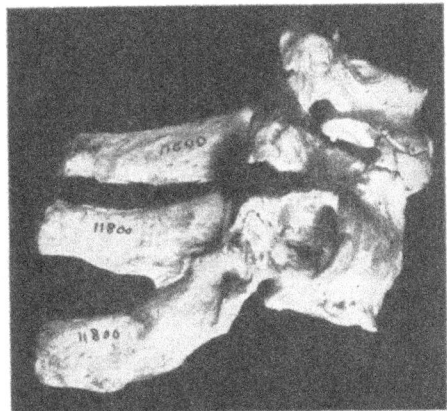

Fig. 7b.

PLATE V

Fig. 8.   Hyaena.   W.I. 7,102.   Page 97.
    A. Left elbow complex showing hyperostosis and ridging of the articulating surfaces.
    B. Left humeral head with hypertrophic arthritis.

Fig. 9.   Hyaena.   W.I. 7,102.   Page 97.
    X-ray of the right elbow complex showing thickening of the bone but no true shadow of the hyperostoses.   A suggestion of ulceration of the internal articulating surface.   Some demineralization of the wall with thinning of cortex.

Fig. 11c.   Leopard.   Elbow complex with marked hyperostotic arthritis.   There are two grades of change in this animal, overgrowth and raggedness suggesting ulceration.

PLATE V

Fig. 8a.

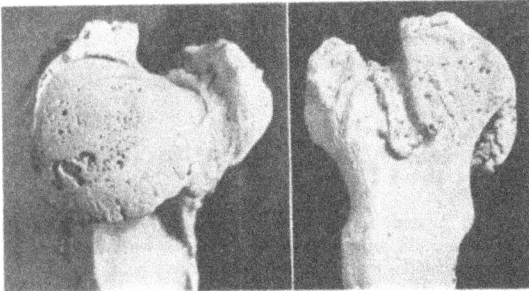

Fig. 8b.

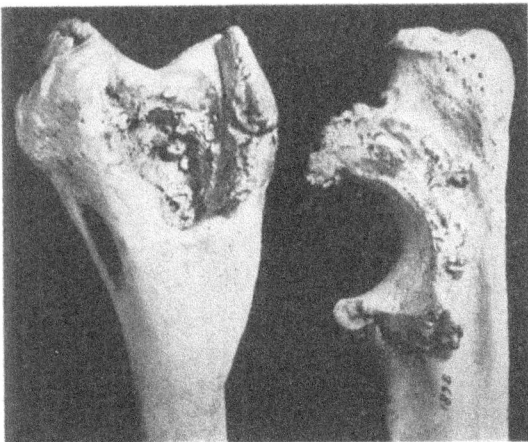

Fig. 11c.

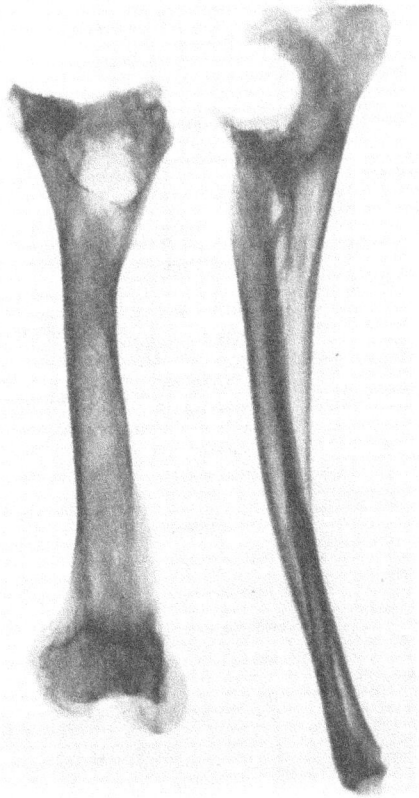

Fig. 9.

PLATE VI

FIG. 10. Photograph of the atlas and axis of a seal. Not included in the bones used for study. Loaned by Dr. Ales Hrdlicka. Found by him in Alaska. Note hypertrophic and erosive condition of both bones.

FIG. 11. Leopard. W.I. 3,681. Page 98.

A. Lumbar spine showing roughness of the articulating faces and calcified bridges.

B. X-ray of the same vertebrae revealing roughness and therefore almost certainly bone ulceration at the edges of the articulating surfaces. Compare this with Fig. 4A.

FIG. 12. Guinea baboon. P.Z.G. 12,105. Page 100.

X-ray of jaws showing peridental thickening around the teeth in the upper jaw, probably unerupted and dislocated molar, scleroses about the upper incisors, absorption of the lower lateral teeth, some porosity at the root of the incisors. This animal had serious spondylitis.

PLATE VI

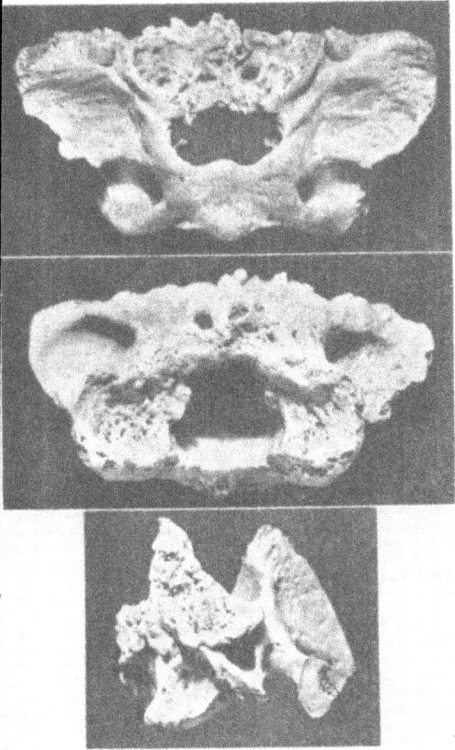

Fig. 10.

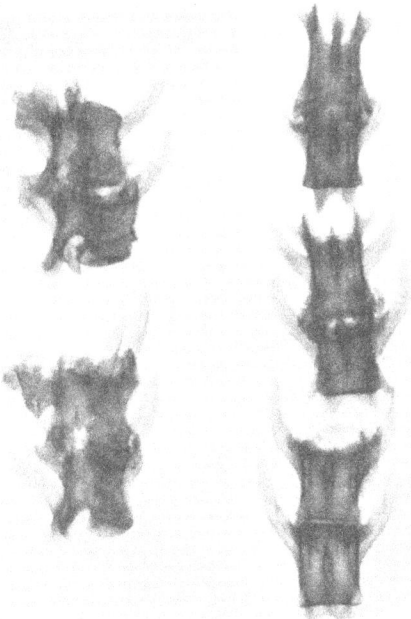

Fig. 12.

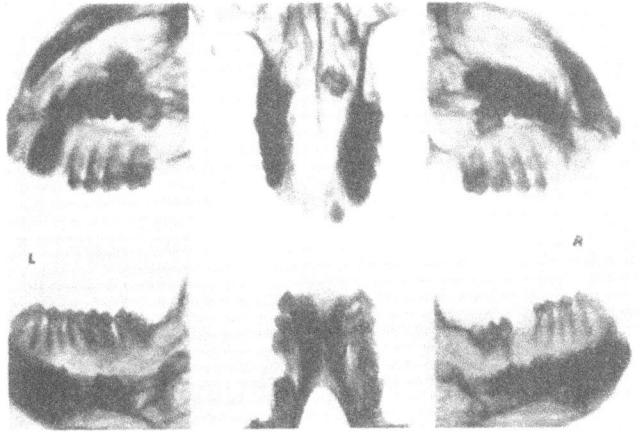

Fig. 11A.

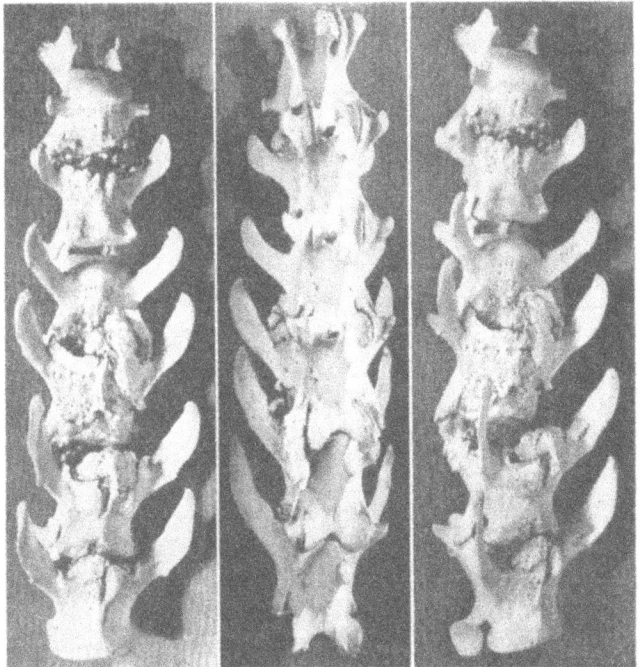

Fig. 11B.

PLATE VII

FIG. 13. Mandrill. P.Z.G. 12,147. Page 102.
   A. Complete spine with four curvatures and advanced destructive spondylitis.
   B. Higher power of the lower cervical and thoracic spine.
   C. Higher power of lumbar spine.
   D. X-ray of spine laterally showing irregularities of the bodies, wedging of certain individual bones, notably 1st, 2nd, 3rd and 5th thoracics, destruction or roughening of the articulating faces.
   E. Longitudinal section of the 5th and 6th cervicals that were completely ankylosed. This section surface shows no trace of the intervertebral articulation of these two units. It is therefore a complete ankylosis with melting together of the two bodies. Compare and contrast with 7B.

PLATE VII

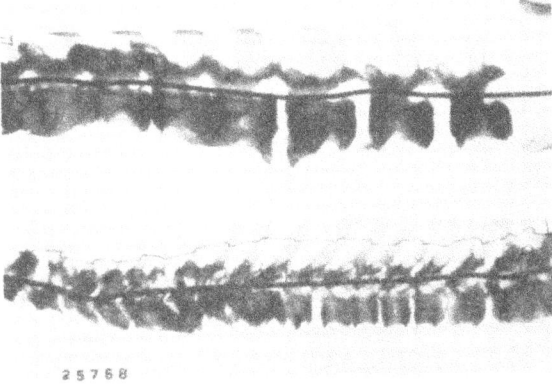

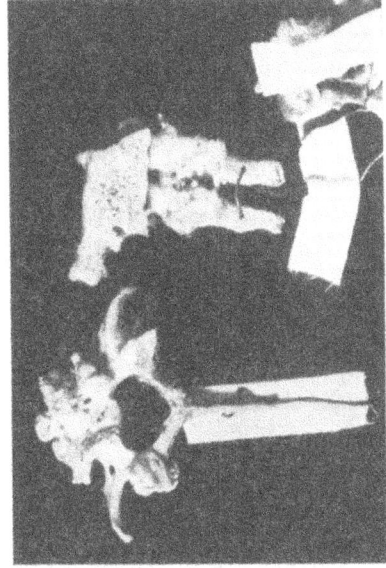

Fig. 13d.

Fig. 13e.

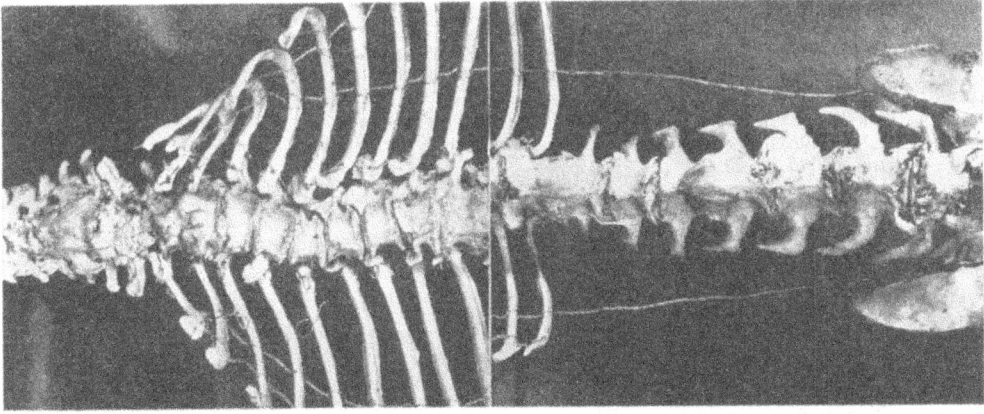

Fig. 13b, c.

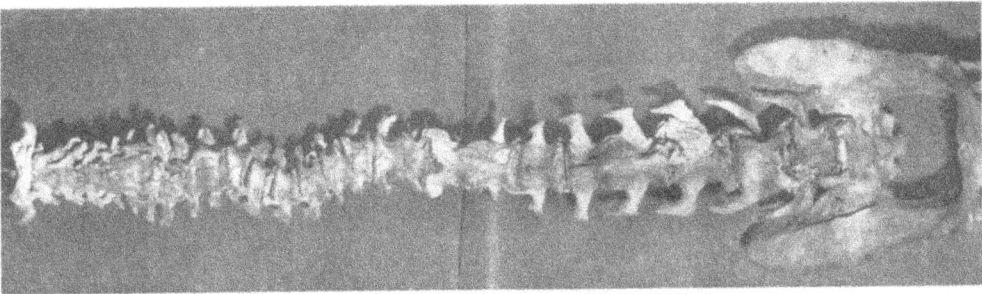

Fig. 13a.

PLATE VIII

FIG. 14. Gorilla. T.D. 1,409, normal.

X-rays (A) right hand, palmar, (B) right foot, palmar. For comparison with other x-rays of gorilla hands and feet.

FIG. 15. Gorilla. T.D. 1,954. Page 105.

A. Right hand, palmar. Marked hyperostosis of distal metacarpals 3 and 4; some on 1. Porosity and irregularity of proximal and distal 1st phalanges 3 and 4.

D. Left knee complex. Eburnation and porosity of the internal condyle; defect of the external condyle is artificial for the observation of the interior of the bone.

E. Left tibiofibular joint, accidentally separated in handling, showing hypertrophic arthritis with true ankylosis.

PLATE VIII

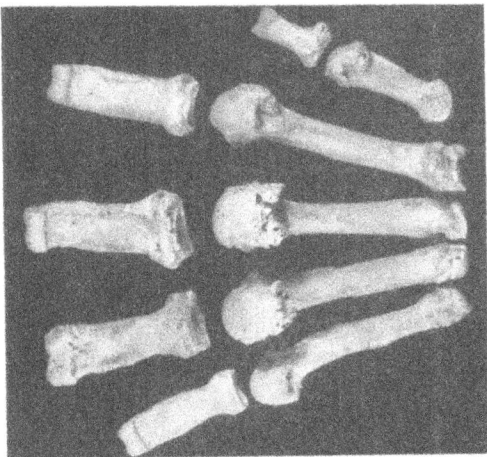

Fig. 15A.

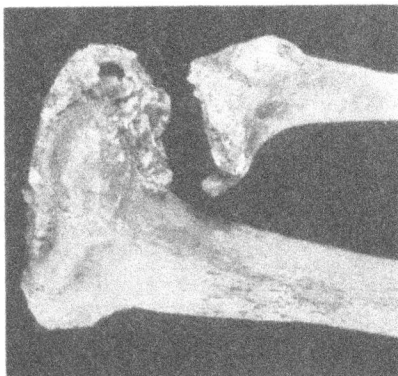

Fig. 15E.

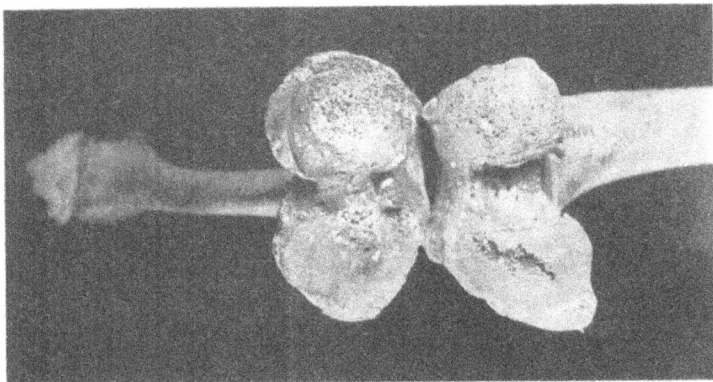

Fig. 15D.

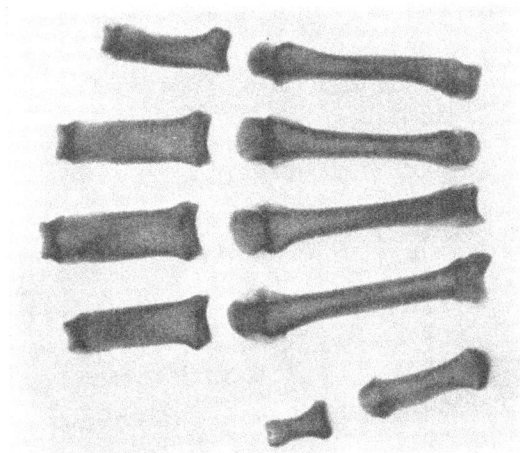

Fig. 14A.

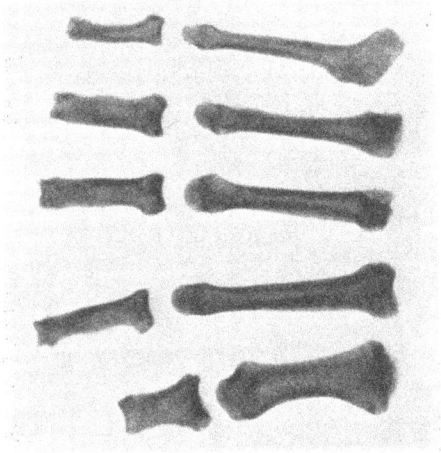

Fig. 14B.

PLATE IX

FIG. 15B. Right hand, palmar, x-ray. Rarefaction of the proximal ends of metacarpals two to five. Striae wide, irregular and obscure. Distal ends hypertrophic with thickening of the epiphyseal lines. Irregularity and porosity in the heads of metacarpals three and four. First phalanges thickening to articular edges with porosities in phalanx three and hyperostosis in phalanx four.

c. Right hand lateral x-ray. Osseous disorganization plainer, especially in destruction of distal ends metacarpals 3 and 4.

FIG. 16. Gorilla. T.D. 1,991. Page 106.

A. Cervicals two to seven, thoracics one to three, first rib internal surface, lipping and porosity of the second and third cervical, marked porosity of the articulating faces of the sixth and seventh. Wedge formation of the first and second thoracic, porosis and hyperostosis of the ends of the first rib.

B. Cervical seventh, thoracic first and second and first rib, lateral right oblique. This is an enlargement of A and illustrates the hyperostotic and the ulcerative character of the change.

c. Lower cervical and upper thoracic, x-ray. Porosity, irregularity of the cortical line and marginal hyperostosis. Wedging of the last cervicals and first thoracics.

PLATE IX

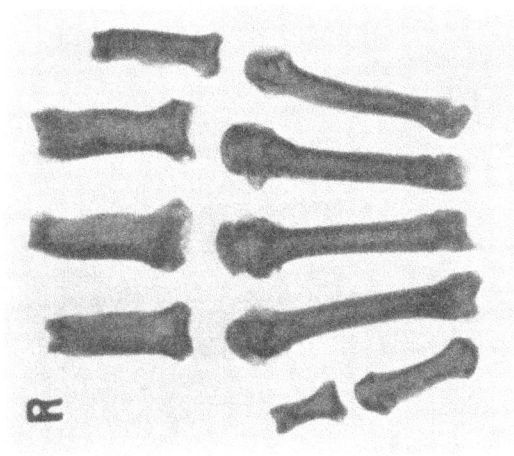

Fig. 16b.

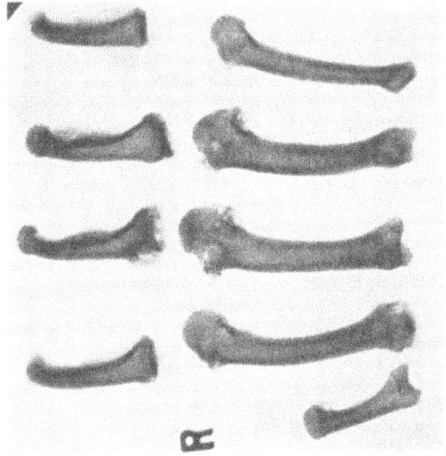

Fig. 16c.

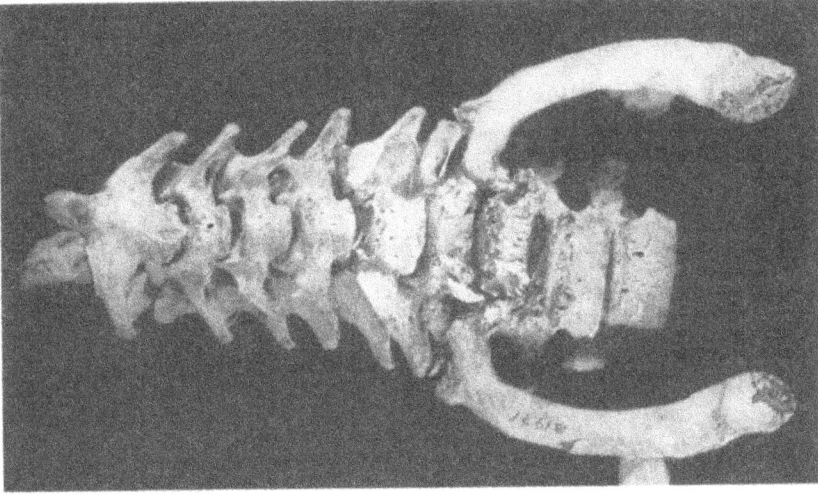

Fig. 16a.

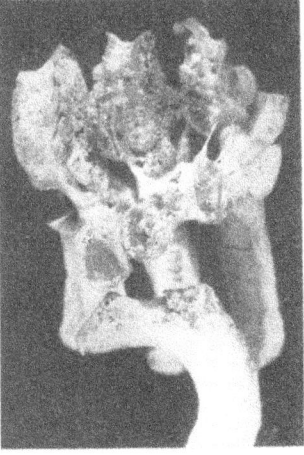

Fig. 15b.

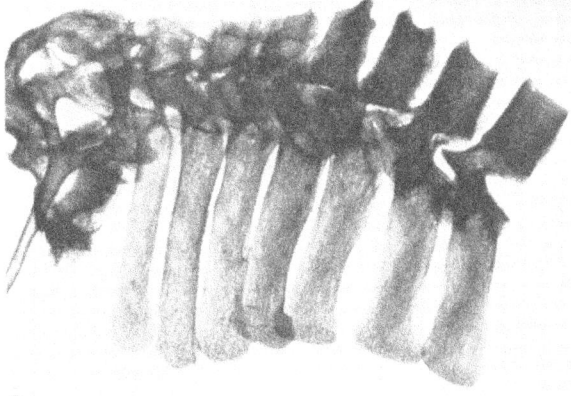

Fig. 15c.

PLATE X

FIG. 16D. Superior surface first thoracic and first rib, x-ray. Porosity and irregularity clearer than in c. Rib head rarefied but surrounded by hyperostoses.

G. Left hand, lateral, x-ray, confirms findings in H. Disorganization very clear.

H. Left hand, palmar, x-ray. Rarefaction and thinning of shafts, striae broken, flattening of the head of metacarpal 4. Marked porosity and rarefaction throughout.

F. Left hand, palmar. Numbers 3, 4 and 5 metacarpal porotic at distal end; 1st phalanx proximal hypertrophic, 2, 3, 4 and 5 all porotic in proximal half.

PLATE X

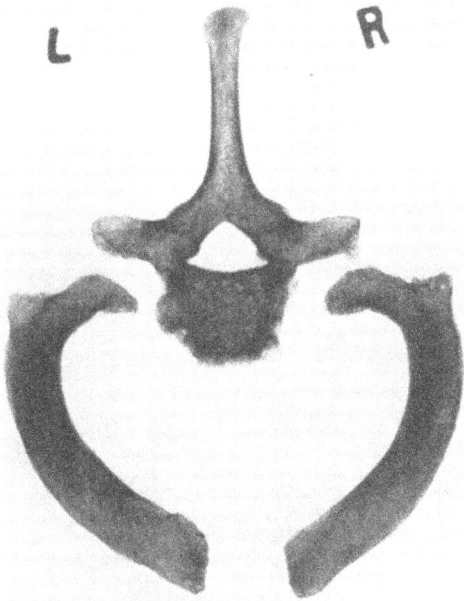

Fig. 16D.

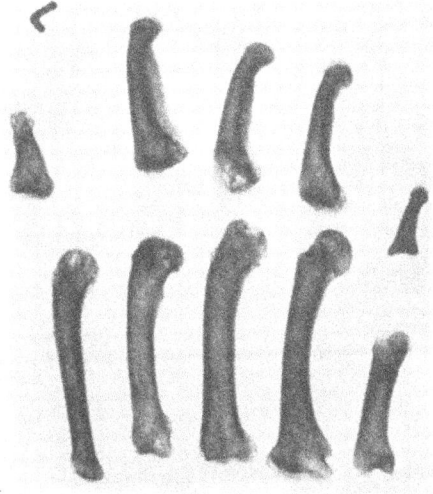

Fig. 16G.

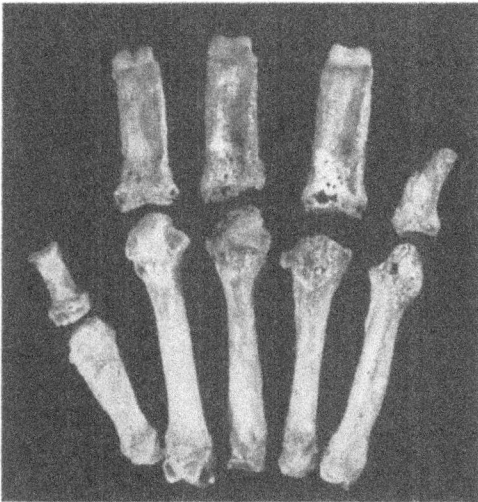

Fig. 16F.

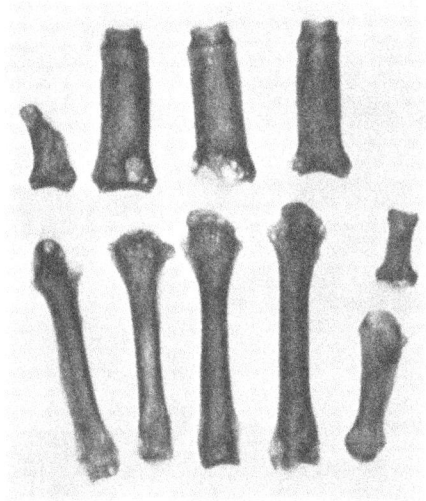

Fig. 16H.

PLATE XI

FIG. 16E. Right astragalus and scaphoid. Eburnation and porosity of surface, there being a distinct subluxation.

J. Left hand, third metacarpal and first phalanx separate, dorsal surface.

K. Second molars, lower jaw, right and left, destruction of the roots, rarefaction of adjacent bone.

FIG. 17. Gorilla. T.D. 1,731. Page 108.

A. Cervical vertebrae, third and fourth. Necrosis and porosity, wedging; the intervertebral discs must have been absent.

C. Right hand palmar. Changes in 4th digit, 1 and 2 phalanxes; early in 5.

D. Right hand palmar, x-ray. Changes in 4th digit; general slight swelling and disorganization of all heads.

PLATE XI

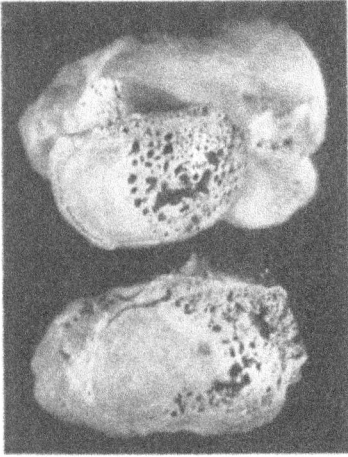

Fig. 16E.

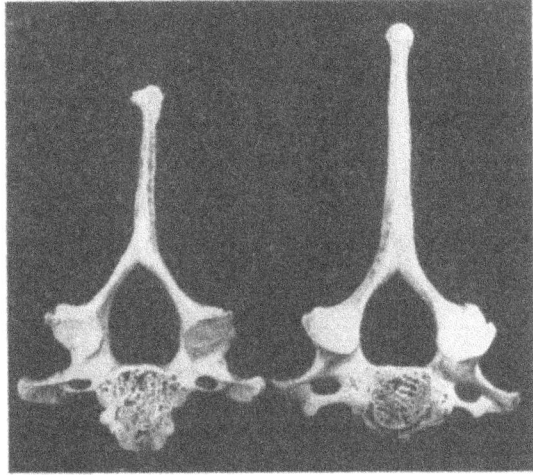

Fig. 17A.

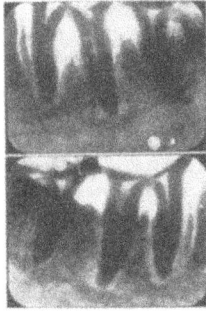

Fig. 16J.

Fig. 16K.

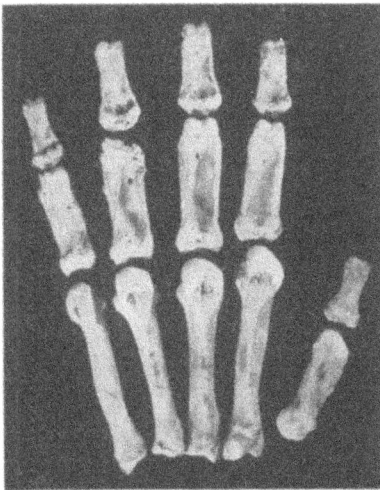

Fig. 17C.

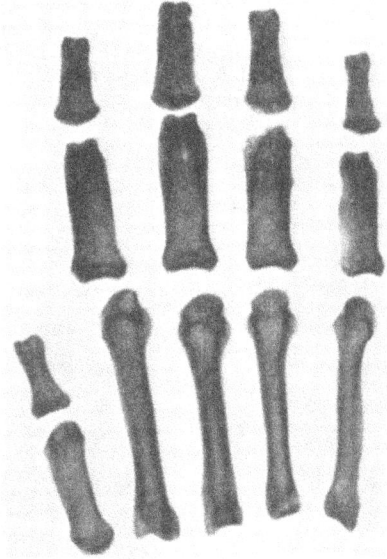

Fig. 17D.

PLATE XII

FIG. 17E. Left hand palmar. Changes show better by x-ray.

F. Left hand palmar, x-ray. Distal ends of the metacarpals are all swollen and demineralized with irregular striae and distortion of articular surfaces, some almost destroyed. Bulging of some bone heads, notably distal of phalanges second and third.

G. Left foot plantar and H, left foot plantar x-ray. Porosity of metatarsal bones, absorption, distortion and rarefaction of metatarsals two to five with hyperostoses and distortions.

PLATE XII

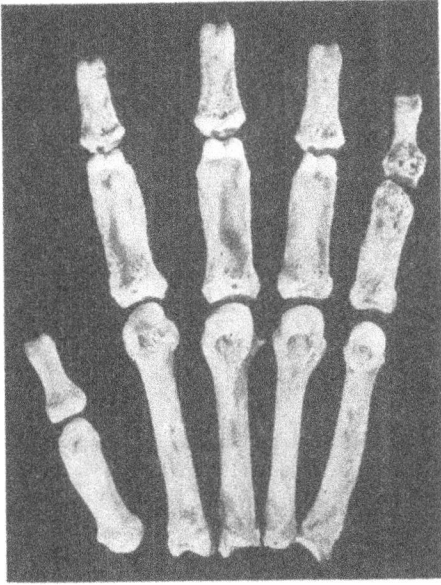

Fig. 17e.

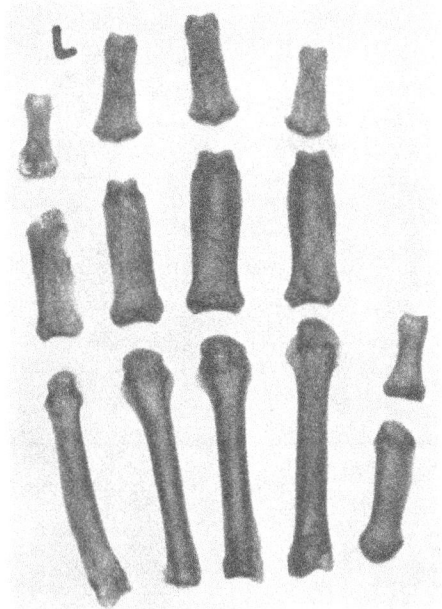

Fig. 17f.

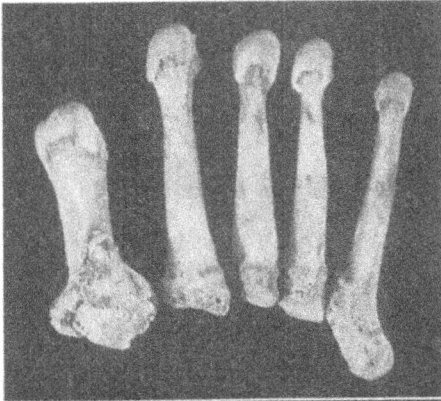

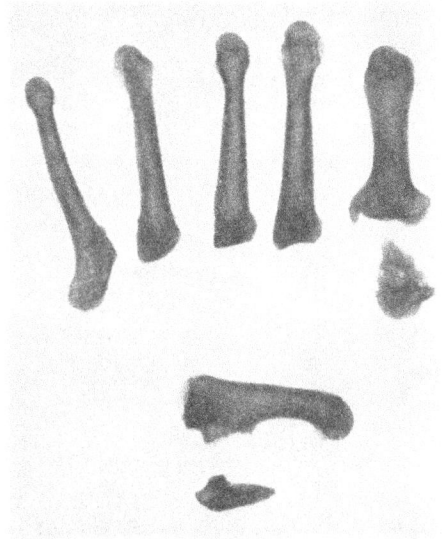

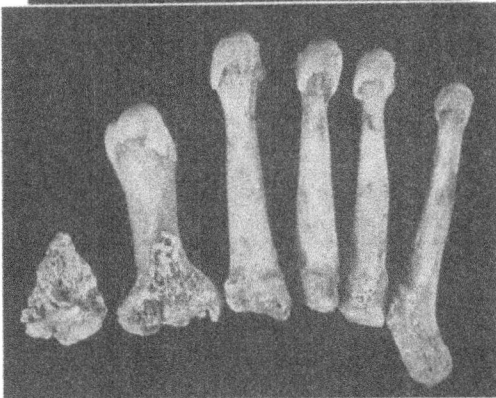

Fig. 17g.

Fig. 17h.